D1085130

HUNGARY
AND THE VICTOR POWERS
1945–1950

HUNGARY
AND THE
VICTOR POWERS
1945–1950

Eric Roman

ST. MARTIN'S PRESS
NEW YORK

ISBN 0-312-15891-2

Library of Congress Cataloging-in-Publication Data

Roman, Eric
  Hungary and the victor powers, 1945-1950 / Eric Roman. — 1st ed.
  p.   cm.
  Includes bibliographical references and index.
  ISBN 0-312-15891-2 (cloth)
  1. Hungary—History—1945-1989. 2. Hungary—Foreign
relations—1945-1989. I. Title.
DB956.R66 1996   95-52046
943.905'3—dc20   CIP

First Edition: July, 1996
10 9 8 7 6 5 4 3 2 1

Contents

Acknowledgments

I wish to express my gratitude to the staffs of the recently opened Hungarian archives—especially Dr. György Lázár of the New Hungarian Central Archives, Mr. László Hubai of the Institute for Political History, and Dr. Béla Pálmány of the Parliamentary Archives—for treating the collections placed in their care with the respect and discretion a national treasure deserves. Their heart is truly in what they are doing and the beneficiary is the researcher, and in the end, hopefully the general reading public.

On a more personal note, I wish to extend my warmest thanks to Eileen and David Hanson, friends always, who were unstinting with their assistance in rendering this bulky manuscript in Word Perfect, and in the process, helped me to become more computer literate.

And it was Ms. Joan Reitz of the Ruth Haas Library at Western Connecticut State University, who, with equal readiness, guided me through the many preliminary phases of the technical execution of my manuscript, and in the end thanked *me* for affording her a glimpse into the history of a fascinating country.

Eric Roman

Preface

The opening of the Hungarian archives for the post-1945 period enables historians—who until now depended largely on "authorized" studies, selectively issued documents, and personal recollections—to gain a broader and clearer understanding of Hungarian domestic and foreign policies in the communist era. The various collections were made available for inspection by parliamentary resolutions, although some archival material, notably that of the ministry of interior, containing files of the infamous security police, are still sealed.

The first declassified archival collection was that of the Institute for Political History (previously Institute for Party History) at 2 Alkotmány Street in Budapest. It contains a treasure house of documents of a semi-official, unofficial, and private nature. In it are the files of the Hungarian Communist Party (MKP), the Social Democratic Party (SDP), and, after the fusion of these two parties in 1948, of the Hungarian Workers' Party (MDP), and of its successor after the Revolution of 1956, the Hungarian Socialist Workers' Party (MSzMP). The extensive files include documents of the Central Committee, of the Secretariat, minutes of the political committee meetings, reports of special commissions, as well as the private correspondence of a number of communist leaders. On the SDP side, of special value are the documents of the Foreign Policy and Nationalities Bureau, which detail the early, forlorn efforts of the Social Democrats to forge an effective policy of their own.

The New Hungarian Central Archives (now integrated into the Hungarian National Archives) is located on Szentháromság tér in Buda. It includes the documents of the foreign ministry—internal memoranda, directives, circulars, two-way correspondence with missions abroad, and documents of the ministry press section as well as the foreign press section. A separate collection contains the journals of the National Assembly and minutes of the parliamentary committee meetings. (There are also a number of more specialized and for the general historians less interesting collections, but their inclusion in this chapter would not be appropriate.)

Considering the relatively short period of time available to the archival personnel (between the parliamentary resolutions and the actual opening of the archives), the material is remarkably well organized. Foreign ministry files are divided by country (with very few exceptions) and each collection is subdivided into "Administrative" and "Secret" designations. The documents are placed in numbered boxes and at times, when thematic reasons call for it, are further divided into envelopes numbered 1/a, 1/b, 1/bc, etcetera. In the Institute for Political History, the documents in each collection are grouped into Fonds, within which "custodial numbers" (örizeti számok) identify the subdivisions.

Advance permission must be obtained from the foreign ministry in order to examine the foreign policy papers at the New Hungarian Central Archives. One should allow at least a month's lead time. Parliamentary journals and parliamentary committee minutes are released at the discretion of the director of the archives. As long as the researcher is duly qualified no difficulty is placed in his or her way. The archival personnel is extremely helpful and reasonably prompt.

Due probably to the paucity of original research material available before 1992, the list of works dealing with Hungarian foreign policy in the immediate postwar years is rather short. No book-length study is available in English. Hungarian Sándor Balogh's *Magyarország Külpolitikája* (Hungarian Foreign Policy) covers the years 1945-1950, but does not avail itself of foreign ministry files. Mr. Balogh has also written books in Hungarian on the history of peacemaking between 1845 and 1947. Mihály Korom produced detailed histories in Hungarian of the events leading up to the armistice of January 20, 1945, but without the benefit of archival material. In English, books of relatively recent vintage include Steven Kertész's *Between Russia and the West: Hungary and the Illusion of Peacemaking, 1945-1947* (1984), essentially a personal recollection, Stanley Max's *The United States, Great Britain and the Sovietization of Hungary, 1945-1947*, (1985) on the western reactions to the subversion of pluralistic democracy in Hungary; and Joseph Kun's brief history of Hungarian foreign policy between 1945 and 1990, in *Hungarian Foreign Policy: The Experiences of a New Democracy, 1945-1992*, in which only a few pages deal with the events covered in this book. In German, there is Heiner Timmermann's *Ungarn Nach 1945* (Hungary after 1945), a collection of essays concerned almost exclusively with domestic affairs.

All of these books are competent and useful but limited in scope and in need of updating. The present book seeks to fill that lacuna and it places Hungary and its foreign policy within the context of the developing Cold War, documenting the gradual subversion of the country's independence in a global struggle that left few small states masters of their own destiny.

Introduction

Through most of its history East Europe, that broad strip of land between the Gulf of Finland and the Balkan foothills, was a reservoir of "submerged nationalities," ethnic groups with scant experience in national independence. For centuries it was under the sway of East European empires such as Austria, Russia, Germany, and Turkey, who kept the region's chronic instability in check. World War I put an abrupt end to this imposed unity. Seven new states emerged in an area where previously there had been but one.* Of these, six bordered on the Soviet Union and three on Germany, each possessing territories on which one or the other power had historic claims. But because Russia and Germany were weakened by defeat and revolution and the principle of national self-determination had a sanctity no power dared to challenge, it did not at first appear that the new states, unstable though they were, faced a threat from east or west. It soon became evident however that independence of the parts meant a ruinous fragmentation of the whole. Strategic wisdom demanded that these states draw together for economic cooperation and defense. But too many ethnic, religious, and territorial conflicts divided

---

*Finland, Estonia, Latvia, and Lithuania had been parts of the Russian Empire; Poland was resurrected from partitions between Russia, Germany, and Austria; Czechoslovakia was pieced together from former holdings of Austria-Hungary; Hungary was a greatly reduced component of that Dual Monarchy; Rumania alone had been previously independent.

them for the formation of a solid, resilient bloc. The French, through a series of fragile alliances, established a *cordon sanitaire* as a rampart against both German expansionism and Bolshevik political penetration. For this cordon to be effective, it was essential that the social and political systems in the region remain stable lest a major upheaval opened the road to revolutionary change. Thus France, and in time Great Britain, gave its tacit support to the semi-feudalistic order that obtained in most of those countries. The result was virtual economic stagnation; industrialization was slow and the pressure on cultivable land increased, creating a large agricultural proletariat.

In the 1930s, under Adolf Hitler's territorial reordering, the cordon crumbled; in 1939–1940 it disappeared altogether. When in 1941 Germany attacked the Soviet Union, most of East Europe fell under its sway and Hitler's plan, outlined some twenty years earlier in *Mein Kampf*, to bring the grain-rich spaces of the Ukraine under German hegemony, came close to realization. But the tide turned after Stalingrad, marking the beginning of a long period of soviet ascendancy. As the Red Army swept into East Europe, it brought with it the revolutionary changes the western powers had tried to prevent. It scarcely mattered that the changes were effected by military imposition rather than internal development. Before the war was even over Josef Stalin built up his own cordon, from the very states with extended perimeters that France had used to build up its own. In due course the conquered nations were forced to adopt the soviet model of governance and economic retrenchment.

Stalin's motives were not entirely ideological or bound up with national security. Territorial expansion, rooted in Russian tradition, as well as Stalin's tyrannical temperament, played a large part. In fact among the victor states, major or minor, the USSR was the only one that emerged from World War II with substantial conquests. What it did not conquer territorially, it in time conquered politically.

Apart from Finland, not intended for a satellite status, only two states in the region escaped instant sovietization: Czechoslovakia and Hungary. Both had been "liberated" by the Red Army, but Czechoslovakia, a "victor" state, was under soviet occupation for only a short time; Hungary on the other hand was saddled with an occupying army of some two million. There was nothing in the nation's past to earn it favorable consideration. In both world wars it had fought on Germany's side and its political sins dated back centuries, especially the opposition to social change of its landed magnates, who dominated the government. Yet Hungary after World War II was allowed to form a multiparty government and maintain it for over three years with only sporadic outside pressure for change.

Hungary was thus an atypical case in East Europe after World War II, as it had been after World War I. In the 1920s it had not been a part of the *cordon sanitaire*; that was a French creation and France allied itself with Hungary's enemies, the so-called Little Entente (Czechoslovakia, Rumania, and Yugoslavia) whose member states had all gained territory from Hungary and drew together against any attempt to revise the treaties. In the years after World War I, Hungary experienced more changes of political fortune than any other nation in the region or indeed in all of Europe. It lost its constituent status in the Austro-Hungarian Dual Monarchy. In 1919 it was the only country outside of Russia to have, albeit for only a short time, a soviet republic. It was the only state in Europe to lose two-thirds of a national territory it had ruled for over a thousand years. It retained the monarchical form of government but expelled the monarch when he tried to return. In the hope of revising the peace treaty it oriented its foreign policy for a decade toward Great Britain. Disappointed in that alignment, in the 1930s it turned to Germany. By 1940 it was the only country in Europe that had more than doubled its territory since the peace settlements. As World War II drew to its close these recently acquired territories were lost again, and the hatreds and enmities engendered by the frequent changes of jurisdiction would blight Hungary's relations with its border states for the rest of the war and beyond. Yet of all the states in the region, Hungary proved most successful in escaping soviet domination and experimenting with democracy on the western model.

This study offers a comprehensive account of Hungarian foreign policy in the five years immediately following World War II. I undertook this work in the face of certain hardened preconceptions. As the sovietization of a country with Hungary's political past, the transformation of an ossified feudal system into a socialistic one, is a remarkable development, most studies of this five-year period have focused on internal history. The foreign policy of a nation under direct military occupation, or even in its sovereign state burdened by a foreign military presence, would appear to have little historic value and would amount to no more than a series of footnotes to the diplomacy of the controlling power. The most a historian can do in this context is to provide a chronological sequence of events dictated from outside.

Yet there were puzzling features about the foreign policy of postwar Hungary from the beginning. In contrast to most of its neighbors, the country had a multiparty coalition government and the conduct of foreign affairs was in the hands of the Smallholders' Party, once a modest agrarian party but now swollen with elements of the political center and right. The control was by no means nominal; members of the left bloc repeatedly complained about their exclusion from the shaping of foreign policy. The Smallholders' Party, with broad initiative, charted an almost openly prowestern course. To the accompaniment of ritual bows to the Soviet Union, it sought relief from internal political and economic pressures by appealing to the western powers, chiefly the United States and Great Britain. While the Hungarian Communist Party assailed the unwisdom of such an orientation, it did little to change it. Not until adverse international developments produced a deep chasm between east and west and a nonaligned position became ever more difficult to preserve did the autonomy of Hungarian diplomacy begin to be abridged. Even then though the process was gradual and halting and brought on more by the psychological pressures of the Cold War than by dictation from Moscow.

Stalin's leniency toward Hungary in these critical years remains a puzzle and none of the explanations are satisfactory. Stalin could not have hoped that a nation which for so long had kept its peasantry in destitution, coined the phrase "a paraszt nem ember" (the peasant is not a human being), and later prevented its nascent industrial proletariat from gaining a voice in the political sphere would reform its antiquated social system of its own will. Nor could he hope that it would turn a genuinely friendly face to its great neighbor in the east. Hungarians instinctively distrusted all things Russian. Nothing but evil had come to them from the east: heathen Cumans and Petchenegs, Mongolians, and Turks, and, in 1849, a large Russian army that crushed the nation's War of Independence against Habsburg rule. The intelligentsia, certainly since the Age of Enlightenment, had looked to the west for inspiration and guidance. In the Dual Monarchy the Hungarians had consistently warned their Austrian partners of the danger of engulfment by the "Slavic sea." Any means of support from the west, even after World War I, was preferable to whatever might have been offered from the east. During World War II, as we shall see, the Hungarian government and the ruling circles spared no effort to effect an armistice with the western powers, to the total exclusion of the Soviet Union.

Stalin was prone to punish lesser sins with the harshest retribution. There were rumors in Hungary of possible annexation to the Soviet Union, of mass deportations, of military governance lasting many years. Yet Stalin asked for little more than that Hungary pay for part of the damage its armies had inflicted on Russia. He allowed it to hold free elections and to organize a broad coalition government; whenever that coalition was threatened, usually by the Hungarian Communist Party, he, or his proconsuls, intervened. Was it tactical patience or was it a piece of a strategic plan?

No final and incontrovertible answer is possible. But Hungary's road from an occupied though politically still free country to a satellite status offers its own lessons and conclusions and amply illustrates the fate of a small nation in a world ruled by great powers.

Chapter 1

The Eleventh Hour

Hungary in the closing months of World War II had two functioning governments in an effective state of war with each other, their spheres of jurisdiction separated only by the fighting front. The government in the west was a carry-over from the country's alliance with Nazi Germany; it remained steadfast in that alliance, never surrendered and was physically driven from the country. Its evacuation was not a mere military retreat; it was comprehensive. In the words of a Hungarian historian of World War II:

> It is without parallel, not only in Hungarian history but in the history of Europe as well, that virtually the entire leadership cadre, the high administrative bureaucracy, the officer corps of the armed forces, the generals and general officers, as well as the police force, should in one body leave the country. It is unprecedented that the entire machinery of the state, its political and economic governance, its parliamentary and law enforcement organs, its defense forces of several hundred thousands, and its [major] institutions, should transfer themselves to another state, sweeping along a significant portion of the social elite, carrying away even the prisoners, certainly the political ones. . . . On German territory, in an illusion of legality, lacking a political home base, this state apparatus lived and carried out its activities for several weeks.[1]

The other government, hastily established in the city of Debrecen, which only weeks earlier had been the scene of savage fighting, drew its legitimacy from a resolution by an obscure grouping of progressive elements and from a more or less implied recognition by the great powers at war with Germany and its allies. The resolution promised to establish a new Hungary, progressive and peace-loving, but its immediate task was to conclude an armistice with the allied powers.

There had been attempts to quit the war for over two years but until the fall of 1944 all overtures had been made to the western powers. The premier in 1942 and 1943, Miklós Kállay, was a large landowner and a prominent member of the agrarian elite, which held it as an article of faith that any peace settlement would have to leave the country's social and political system intact and private property inviolable. Kállay and his political allies understood that these conditions could be achieved only if the country was liberated by Anglo-American forces. The British and the Americans might not have viewed the existing social order in Hungary with great approval but they preferred it to a radical alternative. The Kállay government was even willing to accept the "unconditional surrender" formula enunciated by Winston Churchill and Franklin D. Roosevelt at their Casablanca meeting in January 1943, but only if the surrender was made to the forces of the western powers. A Hungarian "trusted agent," in the summer of 1942, communicated to the British in Istanbul a three-point offer, of which the first and principal point was that "Hungary did not intend to oppose Anglo-American or Polish troops if they reached the Hungarian frontier and advanced into the country. Hungary wished nothing in return for this."[2]

A great deal of Byzantine maneuvering centering on personalities followed and it was not until September 9, 1943 that His Majesty's ambassador in Ankara went to meet a Hungarian emissary on board a British ship in the Sea of Marmora to give him his government's reply. The western allies were of course aware that an acceptance of the Hungarian offer would be tantamount to entering into separate peace negotiations, something that members of the Grand Alliance had vowed not to do. The British note, formally issued "in the name of the United Nations," thus expressly stipulated that "Hungary's capitulation will be kept secret. It will be made public by the Allies and the Hungarian government simultaneously, at a time agreed by both parties as suitable. At the express wish of the Hungarian contracting party it is agreed that the publication shall in no case occur before the Allies have reached the frontiers of Hungary."[3]

Phrased in this manner the precaution might have been against a German coup rather than against soviet meddling, because it referred not to the western allies but to the allied side as a whole. There is no doubt however that the British fully understood the Hungarian motives and were prepared to present the Soviet Union with a fait accompli. Budapest took the prospects of an early peace so seriously that the foreign ministry organized a special section to prepare terms for a future peace conference. The group continued its work until the German occupation of the country in March 1944, when it was disbanded.[4]

At the time of these contacts Churchill's plan of opening a second front in the south of Europe rather than on its western coast was still alive; when a British-American force landed in Algeria early in 1943, took Tunis, and later invaded Sicily and Italy proper, the Budapest government concluded that the western powers sought to make the Mediterranean the main theater of operations. The Red Army was still battling deep inside Russia and for an eventual invasion of Central and East Europe the western powers had the positional advantage. The Hungarian government, in line with its earlier promises to the British, sought to reduce the scale of its cooperation with the Germans. When at the end of August 1943 Berlin demanded that Hungary send occupation troops into Yugoslavia to relieve German forces for combat duty, the Hungarian deputy foreign minister wrote to the chief of the general staff: "The Anglo-Saxon powers will sooner or later appear in the Balkan region," and voiced strong opposition to complying with the German request.[5]

At a meeting of the allied leaders in Teheran later that year Churchill's plan of advancing into Europe through the Balkan peninsula was set aside in favor of a cross-Channel invasion. That winter units of the Red Army came within striking distance of the Carpathians and prospects of a separate peace with the Anglo-Saxons quickly faded. Alarm gripped the Hungarian ruling circles. However, they still could not bring themselves to believe that the western powers would meekly surrender East Europe to the soviets. A plan was prepared in the inner councils of the government and communicated to the American Office of Strategic Services (OSS) which in turn forwarded it to the State Department. The OSS officer in charge of these contacts reported that the Hungarians "would be willing to risk a break with the Reich to try to defend their own frontiers if safeguards could be obtained against an invasion by the Russians. He [the Hungarian contact person] believes that an occupation by the Anglo-Saxons would be welcomed by Hungary."[6]

As late as March 1, 1944, when Kállay composed his report on the international situation, he asserted that his government based its policy on the assumption that "although the Anglo-American troops are far from Central Europe, their policy still reaches us. We thus assume that the Anglo-American forces will reach this area ahead of the Russian soldiers and that Anglo-American policy will decide the future of East Central Europe."[7]

When the essentials of this report were received in London and Washington, a State Department official summed up his impression of it in these words: "The Hungarians now say that [they] are not an enemy of the United States or Great Britain and will not offer resistance to Anglo-American troops, that the Hungarian Government adheres to the Atlantic Charter and trusts the wisdom and motives of the American and British Governments and that the present Hungarian attitude is serving the Allied cause in many respects, but that the whole nation will rise and defend itself against a Russian attack."[8]

Little sympathy was forthcoming from the western capitals, however. Allied unity was still unbroken, Hungary was an enemy state and, the American position continued to be that "We don't deal with any of these overtures except on the basis of unconditional surrender."[9] Budapest was advised to turn with its pleas for an armistice to Moscow.

That was precisely what the Hungarian government had tried to avoid in its search for a way out of the war. The head of state, Miklós Horthy, a former admiral who since 1921 had acted in the name of the absent monarch Otto von Habsburg and had appointed the country's successive governments, had made anti-communism the cornerstone of his internal and external policies. An appeal to the soviets was for him and those around him a most unpalatable option. Three years earlier Horthy had enthusiastically endorsed Hitler's "crusade" against soviet Russia. He had written to the Führer that "when we think of the future of western culture, which has been built up by the heroes of mankind over millennia, we cannot escape the feeling that the greatest danger that threatens us all is Red communism. It is the aim of communism to annihilate culture and subject the world to its rule."[10]

Now the destroyer of cultures stood at the gate. On August 22, 1944 the Red Army breached the easternmost frontier of the enlarged Hungary in the Carpathians. Still the hope of an Anglo-Saxon presence that would rescue the country from soviet domination would not die. At a meeting of the ministerial council on August 25 one participator ventured that "we have to reckon with the possibility that the English will attempt a landing on the Adriatic coast . . . [hence] we should hold back the Russians until the Anglo-Saxons have been able to occupy Hungary."[11]

These frantic and single-minded efforts rested on deep convictions Kállay later poignantly expressed in his memoirs:

> It is undeniable that Hungary made her offer [to the western powers] not only with the object of escaping from Hitler's grip and taking part in the struggle against him, but also of securing her independence

against all aggressors and primarily against the Russians. Germany could suddenly crush us at any moment, it was true. That was the nearest danger in terms of time and distance that threatened our independence, and somehow we hoped that the agreement [with Britain and the United States] would help us here also. But the first task and object of the Hungarian Government was to protect the nation by linking up with the Allies in order to avert the even greater devastation which threatened us from Russia. We should certainly have been able to free ourselves by our own forces from the Germans, but not from the Russians. The only Trojan horse that the Nazis could send into our city was anti-Semitism, that is, agitation against one group. The weapon of the Russians, however, was the general war and beyond that war against everyone, irrespective of class, who did not bow the knee and the head not only to Bolshevism but also to the biggest imperialism in the last five centuries. One of the fundamental and catastrophic anomalies in the situation was that we were seeking the protection of the British and Americans, not only against our own "allies," but against their allies also.[12]

Hopes of escaping all the evil that the very name "Russian" conjured up in Hungarian minds suffered a fresh blow when the Rumanian government, which had fought on Germany's side since the start of the Russian campaign (and was, by perverse irony, an ally of Hungary too), announced that it was withdrawing from the war. On August 23 the king, after dismissing the pro-Nazi head of the government, General Ion Antonescu, declared war on Germany. Horthy used the occasion, and German preoccupation with the loss of an ally, to dismiss his own pro-Nazi premier and former ambassador to Berlin, Döme Sztojay, whom the Germans had foisted on him after they occupied the country, to appoint a military man with liberal leanings in his stead.*

For Hungary, Rumania's desertion was of the greatest military and political significance. It served as a reminder that the unconditional surrender formula had variants and a German ally, even when it had fought against the "United Nations" for years, could still establish its bona fides and bid for favorable consideration after the war if it renounced its alliance and declared war on the common enemy. On the military side the road was now open to the Red Army to race through Rumania's undefended plains and enter Hungarian Transylvania from the south as well as the east.

Under the impact of these events Horthy at last decided to do the unthinkable and appeal for an armistice to the soviets. After two tentative and unsuccessful attempts he established contact with Moscow at the end of

---

*Premier Kállay had been forced to resign prior to Sztojay's appointment and he took refuge in the Turkish embassy. When the Germans pressed for his extradition, he left the embassy voluntarily and was taken into custody.

September. Earlier approaches had been made not directly to the soviet government but to the soviet partisan command in Slovakia; they produced a curious and still controversial document, a letter from the head of the partisan command, Makarov, purportedly spelling out the conditions under which Hungary would be granted an armistice. The letter clearly intended to dispel fears of what occupation by the Red Army might mean politically, for it contained assurances of a remarkably liberal nature. It promised that "Hungary's complete [territorial] integrity will be guaranteed and in case a soviet-Rumanian agreement temporarily assigns [all of] Transylvania to the Rumanians, in the peace negotiations the Soviet Union will represent the principle of a plebiscite for the territory of Transylvania." It further stated that the Rumanian troops who, as allies of the soviet forces, were advancing on Hungarian territory would be halted and only the Red Army would move forward. Thus Hungary's independence would be assured. "She can choose her own form of government and the Soviet Union will not exert internal political pressure on the Hungarian regime." The letter further promised that a Hungarian armistice delegation would in Moscow negotiate not only with the soviet government but with "all three great powers." Finally, "the Soviet Union does not wish to expand in Europe."[13]

Although the contacts with the partisan command were informal, they committed the Budapest government to negotiating in Moscow, primarily with the soviet side, and earlier plans formulated by the British Foreign Office with American concurrence were no longer timely.* The purpose now was to gain reasonably fair terms before the soviet government demanded unconditional surrender. A delegation headed by the inspector general of the gendarmerie and including the regent's oldest son was dispatched and reached Moscow by a roundabout route. It carried a letter from Horthy to Stalin, asking for an armistice but stressing the need for Anglo-American troops to participate in the occupation of Hungary. The letter also asked for safe conduct for German forces withdrawing from Hungary.[14]

The soviet negotiators either did not know of the Makarov letter or looked on it as passé, because when Foreign Minister Vyacheslav Molotov handed the Hungarian delegation the "preliminary" armistice terms on the night of October 8, they contained no concessions or any of the assurances proffered by the

---

*Those plans envisioned armistice talks to be held in Italy, or possibly in Turkey, with soviet participation. The British had located a Hungarian general in Italy who, although not suitably empowered, had sufficient authority to sign an armistice agreement.—*Foreign Relations of the United States* (Hereafter, FRUS), Washington, D.C.: Government Printing Office, 1944 Vol. III, 887-89.

Makarov command. Instead they demanded withdrawal, within ten days, of all military and administrative personnel from territories Hungary had occupied as a result of Hitlerite arbitration, the acceptance by Hungary of an interallied control commission, and an immediate declaration of war on Germany.

The soviets were leading from strength. Their forces, after the Rumanian surrender, were sweeping through Transylvania practically unopposed; on September 23 units of the Second Ukrainian Front had entered Hungary proper. Three days later they occupied the first major town in Trianon Hungary, Mako. They stood at the southern edge of the Great Hungarian Plain, which offered no major obstacles to infantry or armor. Hungary's future hung in the balance and Horthy had little time to lose.

On October 15 he announced over the airwaves that Hungary had withdrawn from the war; he ordered his armed forces to cease fighting and resist German attempts to disarm them. But Horthy had tarried too long. The Germans had anticipated the move for some time and enlisted the alliance of the Arrowcross Party, Hungarian Nazis with very slim popular support. The party leader, Ferenc Szálasi, and the narrow-minded, mediocre people who formed his entourage either did not realize the hopelessness of the military situation or else they were ready to go down to defeat for the sake of an ephemeral possession of power. Szálasi appointed himself *nemzetvezető* (Nation Leader) and in a radio broadcast denounced Horthy's attempt to desert Hungary's ally and called his order to stop fighting against the Bolsheviks treacherous. The Germans took Horthy prisoner and deported him with members of his family to Germany. There began five months of Arrowcross rule that witnessed the deportation of some 40,000 Jews (in addition to the approximately 400,000 who had been shipped to German death camps in the spring) and turned the country into a battleground. The wealth of the nation was plundered, the bulk of its movable assets was transferred to Germany, and Hungary became the last holdout in the Axis camp.

The soviet military command now faced the dual and conflicting task of waging all-out war on Hungarian territory while convincing the populace that the Red Army had come as friend and not as enemy. The military effort was uneven, producing some hard-fought battles and some rapid breakthroughs. Reassuring the public was a more difficult task. Alarming reports had been coming from Transylvania of the savage conduct of soviet soldiers, of rape and pillage; these reports only confirmed the propaganda of the past two decades that portrayed Red Army men as Asiatic brutes with no respect for the rights of civilians or for private property.

To counteract these fears soviet authorities embarked on a series of "informational" meetings for the enlightenment of both soldiers and the population in the conquered towns and villages. The initiative came from the command of the 53rd Army of the Second Ukrainian Front, which had been the first to enter Trianon Hungary from the southeast. At the meetings for Red Army men the speakers explained that the nation which it was now their duty to conquer had a glorious past, distinguished by many noble struggles for progressive causes.

The populace on its part was assured that soviet soldiers had come to Hungary as selfless liberators; the oppressive rule of large landowners and of a gendarmerie that guarded their bloated wealth was over.

A widely distributed leaflet featured the picture of an old peasant at the side of a young woman holding a sheaf of grain under her arm against the background of a city in whose skyline a church spire and the tall smoke-stacks of a factory, belching smoke, were prominent. The legend read: "Hungarians! The Red Army comes to you not as conquerors but as an army that frees you from the yoke of German oppressors."[15]

These attempts were reinforced by the issuance of a multipage pamphlet reminding the soviet forces that Hungarians had for four hundred years struggled against German oppression. It detailed phases of that struggle and furnished the impressive statistic that in the year 1917 some 200,000 Hungarian prisoners of war had joined the communist cause in Russia. In time-honored fashion the pamphlet drew a distinction between the laboring people and their fascist rulers. A closing proclamation admonished the soldiers to conduct themselves in a manner that would honor their own traditions and those of the Hungarian people.[16]

Amid all this uncertainty and the attempts to ensure calm one thing remained clear to the populace and the occupiers alike: the Hungarian Communist Party (MKP) that for a quarter century had lived (when it existed at all) underground and had given the laboring people scores of martyrs, if not heroes, would become the most consequential force in the nation's political life. The leaders, who had been exiled or exiled themselves to the USSR, had not yet returned. There were faint stirrings among domestic communists after the arrival of the soviet forces; just how effective their actions were was hard to tell. Postwar propaganda stressed the leading role of communists in every prodemocratic and anti-fascist initiative, but the initiators remained largely nameless.

There had come about under German occupation a loose confederation of center and left parties that called itself the National Independence Front; its chief spokesmen came from Social Democratic and

Smallholder ranks.* On September 20, 1944 the front, in a formal memorandum to Regent Horthy, had demanded that Hungary end its participation in the war on Germany's side and that German forces on Hungarian soil be disarmed. At about the same time underground communists drew up a program of their own (which was preserved in the archives but until now had not been published), beginning with the sentence: "Horthy must go." It then demanded that both houses of parliament be dissolved and that even lower-level assemblies, "which in their composition are hallmarks of reaction," be terminated and the gendarmerie disbanded. "In its place a national guard must be formed from among workers, peasants and the intelligentsia as well as the petty bourgeoisie." There were to be general, secret, equal elections, with women participating. The document further demanded freedom of the press, assembly, and religion, and all the other freedoms that had been enshrined by democratic ideals.[17] A similar memorandum was drafted by Hungarian communists in Moscow, typed on the stationery of the party press organ; it included demands for the eight-hour workday, the protection of small landowners from encroachments by large estates, paid vacations for employees, and the regulation of female and child labor.[18]

The expatriate communists in Moscow had not long ago been targets of Stalin's worst purges, but during the war they had gained a measure of security and respectability. Stalin, ever the pragmatist, recognized the opportunity to groom these denaturalized comrades for leadership positions in the countries the Red Army would conquer in the course of the war. They were thoroughly indoctrinated, not only in Marxist theory but also in the political and police methods Stalin had perfected during his long and difficult struggle against the opposition in the Soviet Union.

Most prominent of the Hungarian emigrants, the recognized first among the equals, was Mátyás Rákosi, the oldest son of a Jewish village grocer, a man of small, almost dwarfish, stature but of a massive build and a capacious mind. From his early youth he had devoted himself to revolutionary work with great intensity and an indifference to his personal fate. He had spent 18 years in Horthy's prisons; in 1940 he was exchanged for some battle flags captured by the Russians in Hungary's ill-fated War of Independence in 1849. Others were Ernö Gerö, an economist and trained Marxist theoretician, a grim and humorless man; József Révai, the quintessential communist littérateur, eloquent in speech as well as writing; Imre Nagy, a former locksmith and later an agrarian economist who, as a prisoner of

---

*The Smallholders' Party (KGP) membership was composed largely of holders of medium-size farms and of petit bourgeois; after the war, with the disappearance of the rightist parties, the KGP attracted many Horthyist "reactionaries."

war, had joined the Red Army in 1917 and had since served international communism in various posts and places; and Zoltán Vas, another old communist, the first to return to Trianon Hungary when the southeastern corner of the country fell to the Red Army.

Vas's hastily written letters to his comrades in Moscow provide a vivid picture of life in these early days of soviet occupation. The dominant theme was the outrageous behavior of soviet soldiers, which the authorities were unable to check, despite all kinds of countermeasures, including executions. "Talked to officer corps of several divisions," Vas wrote in one of his letters, "and explained to them that these atrocities make our [communist] work impossible. Worse even is the panicky mood of the populace. People don't dare to leave their apartments, in terror of the soldiery. . . . Economy wretched. Requisition of cattle and hogs widespread. Everything has in effect passed into the hands of the Red Army. They even took away the fire engines. No commerce, shops plundered."

A few days later he wrote again, from Szeged, the second-largest city in Hungary. "The city has no government, the communists have not come out from underground. Whoever appears in the street is driven away for forced labor. I requisitioned a party building and hopefully now comrades will emerge. But . . . I can't carry on alone. . . . Oh, how we need Révai! And you, Comrade Rákosi, come, come, let's do this together."

Vas repeated his plea some days later, on October 25. "I absolutely need Rákosi. . . . Problems are grave, we need his prestige. . . . Decisions will have to be made on the spot because we lack direct connection to Moscow." Vas then refers to "an added curse," the so-called Seventh Section, a fanatical, ultra-radical faction of the party, and calls its members "stupid, stupid, stupid." He adds: "They ban the red-white-green [Hungary's national colors], they ban the only newspaper in Békéscsaba [a major town in southeast Hungary] because it is published by the Smallholders; they say only communist papers can be published. Then they close those down because they find fault with them. We'll have to turn to Stalin to curb them." Then Vas raises the soul-wrenching question. "Are the people friend or enemy? We must assume friend but [they look on us] as enemy. And now those accursed Rumanians have come to Hungary. They drive everything away, even the men. Only Stalin can help." The letter closes with "affectionate comradely greetings to Rákosi" and adds, "Révai, you come too."[19]

There is no trace in Vas's letters, or in those of his comrades, of the theme that had carried Bolsheviks through the frightful ordeal of the

Civil War a quarter century earlier; that they were passing through the labor pains of a revolution that would, when accomplished, lift mankind to new and glorious heights from which it would never look back again. Many of the illusions of 1919 were broken, communism was no longer a novel experiment, and the sacrifices it demanded were even harder to bear. How was the nation to be convinced that a system that produced this kind of a soldiery, which condoned rape and plunder (while it proclaimed its admiration for the people it conquered), was the vanguard of social and cultural regeneration?

By early November 1944 the two great soviet military formations in Hungary, the Second and Fourth Ukrainian Fronts, had reached the line of the Tisza River that bisects the Great Hungarian Plain. In mid-November the Third Ukrainian Front, having crossed the lower course of the Danube and its tributary, the Száva River, began its momentous drive northward. On December 9 it linked up with units of the Second Ukrainian Front fifty kilometers south of Budapest. The siege of the capital had in effect begun.

"Nation Leader" Szálasi, ensconced in the royal castle in Buda, was making plans to transfer his government to the western part of the country, Transdanubia. During his almost daily visits to Hitler's emissary, Edmund Veesenmayer, the latter transmitted advice and instructions from the German foreign ministry. Veesenmayer's chief task was to put heart and steel into Hungarian resistance and to keep Szálasi from faltering. When he heard of the plan to relocate the government, he strenuously objected. He wrote to Foreign Minister Joachim von Ribbentrop: "In view of the completely inadequate preparation, but in the first place because of psychological and practical considerations, I judge the transfer of government to be mistaken at a time when its imperative necessity is not justified by the military situation." He prevailed on Szálasi to shelve his plan, but only after issuing the categorical assurance that the German legation would remain in Budapest, "even if the Hungarian leaders transfer themselves to the west."[20]

A week later, with the Russians closing in, Szálasi, according to Veesenmayer's report, admitted that "his present position is by no means popular [as] . . . he took over a legacy that could be compared to a festering abscess. He cannot improve on the situation, if only because the Hungarian front is, unfortunately, in constant retreat."[21]

Ribbentrop's daily instructions from Berlin centered almost exclusively on the need to remove the approximately 400,000 Jews still left in Hungary. His stated rationale was the desperate need for labor service in the Reich, but more likely Ribbentrop realized that time was running out on the "final solution" of the Jewish question, and the presence of a large bloc of Jews in a state adjoining Germany offended his racial sensibilities. Several attempts were made to ascertain how many Hungarian Jews were still alive. Veesenmayer on October 28 estimated that at the time the Germans occupied Hungary the total had been "roundly800,000." Since then: "Shipped to Reich territory [almost all to Auschwitz]: 430,000; in Jewish labor service:* 150,000; in Budapest: 200,000.[22]

Szálasi and his clique, by their servile cooperation with the Germans, were sacrificing Hungary's postwar future for the causes of racial purity and the fending off of the Bolshevik menace. The populace in the part of the country controlled by the Arrowcross gave no sign of its disapproval of such suicidal policy. There was no resistance to this doomed and in every sense illegitimate regime. The armed forces by and large obeyed Szálasi's commands and German and Hungarian units fought side-by-side until the last of them were driven from the country on April 4, 1945.

Attempts to create an alternative government that might yet salvage Hungary's future and present itself as the authentic representative of the nation had begun while the capital was still under siege. The armistice delegation Horthy had sent to Moscow was still there, waiting to resume the foundered talks. But these men, all members of the old regime, had become largely irrelevant. The conscience of the nation had to speak with a new voice.

---

*Jewish men were not allowed into the regular armed forces. They were organized into labor brigades with tasks ranging from mine clearance on the eastern front to the digging of drainage ditches, the felling of trees, and the crushing of stones.

Chapter 2

Leaving the War

The proclamation of the "new" Hungary implied the resolute renunciation of the old. The term most commonly used in reference to that old Hungary was "reaction," and it was well-suited for Marxist terminology. Communists portrayed themselves as a party of action, building a new egalitarian society in place of privilege; all those who opposed them were in the service of reaction. The term applied to a variety of elements, mainstays of the old order: the titled nobility; large landowners; monarchists; the Church, Catholic, Lutheran, and Calvinist; the high and middle officialdom; the gendarmerie, which for centuries had zealously guarded the rights of the propertied classes in the provinces and had incurred the additional odium of carrying out the mass deportation of Jews; and all those who after World War I agitated for territorial revision.

Any association with the past regime became a political liability. The future belonged to the progressives. It was a label that every party and every group with political ambitions eagerly adopted. Even the Szálasi regime in western Hungary tried to pose as progressive by commingling the concepts of nationalism and populism and championing the cause of the poor. But the time for such political experimentation was past. Military defeat had crushed its legitimacy; the field belonged to those who had earned their political credentials in the service of the international workers' movement.

But the communists who emerged from underground lacked maturity; some agitated for immediate sovietization while others were content mouthing undigested Marxist phrases. Only the Moscow-bred cadre retained a realistic perspective and could be counted on to make the party a responsible political force. During November two more "Moscow communists" arrived: Ernö Gerö and Imre Nagy. They came with instructions to proceed with restraint. Together with some domestic communists they were instrumental in forming "people's commissions" appointed to fill the gaps left by the exodus of thousands of local administrators. The Hungarian Communist Party (MKP) was officially formed in this month in Szeged; it was, for all its heroic underground past, a new political presence. Unlike its sister party, the Social Democratic Party (SDP), it had never participated in elections, had never had a press organ, and had been banned from agitating; many of its members had been imprisoned and tortured for violating that ban. The nimbus of martyrdom clung to even those comrades who had survived without scars.

For now the party opted for coalition politics. The so-called National Independence Front was in many ways its creation and included, in addition to communists, the Smallholders; the Social Democrats; the Citizens' Democratic Party made up largely of middle-of-the-road intellectuals; and the National Peasant Party (PP), which served as the rural arm of the MKP. The front undertook the task of leading Hungary out of the war without forfeiting its independence.

The armistice conditions Regent Horthy had communicated to Stalin were by now out of date, especially the one that called for western participation in the occupation of Hungary; the most Hungarians could hope for was that the British and the Americans would have an equal voice on the Allied Control Commission (ACC) that would be formed to ensure compliance with the armistice terms. Inasmuch as the Szálasi government was totally unacceptable to the allies for negotiating purposes and in any case would not ask for an armistice, a new government had to be formed in the liberated part of the country; the soviets alone could authorize such an action. A delegation was accordingly sent to Moscow; it consisted of the two communists who had only recently returned from the soviet capital, Gerö and Nagy. They departed at the end of November and began their negotiations on the first day of December. Simultaneously with their departure the National Independence Front composed a solemn plea to all three members of the Grand Alliance. It asked them to grant Hungary the privilege of taking its destiny into its own hands. The true motive behind the appeal was to keep the western allies abreast of developments and thus ensure that the USSR would not be free to set its own conditions.

Somewhat to the discomfiture of the Gerö-Nagy delegation, the Kremlin was just then playing host to General Charles de Gaulle who had come with his foreign minister to sound the soviet leaders on the kind of government they would prefer to see in France. Thus the Hungarians were able to have Molotov's undivided attention for only two days, on December 1 and 5. Molotov was ably assisted by the head of the European section of the foreign ministry, Vladimir G. Dekanozov. According to the notes taken by Gerö, Stalin on the first day occasionally sauntered into Molotov's study where the negotiations were being held and offered his advice, which naturally amounted to an order. Gerö asked for permission to establish in Hungary without delay a representative government. Dekanozov advocated a national advisory commission attached to a government in the hands of the occupiers. Stalin sided with Gerö. But he understood the problem posed by the existence of a separate and functioning government in the west. "They will ask," he said, "whence did [the new government about to be formed] come? We would have accepted Horthy but the Germans took him away. They forced him to sign a document [transferring state authority to Ferenc Szálasi]. Once you have a document it doesn't matter how it came about. Horthy is a moral corpse. Szálasi has something in his hands. The generals [Horthy's emissaries] have nothing. We have to create a source of authority."[1]

To the surprise of everybody present, Stalin insisted that the new government should include Horthy's peace-seeking generals. They would be the ones who would have to accept the conditions Stalin had in mind. "They have to be told everything frankly," Gerö quoted Stalin as saying. "Frank politics is the best politics. They have to be told that the borders will be the old ones. They have to be told that they will have to pay. They have to be told that if they refuse [to participate in the government], others will come and there will be a leftist government."[2]

The conferees agreed that the sovereignty of the new government should stem from a national assembly elected through local organizations, cooperatives, commissions (as had been set up for local administrative purposes), and such other bodies as had popular support. "If there are trade unions," Stalin said, "they definitely have to be included. . . . The party, for now, not."[3]

When the talk turned to the composition of the executive branch, Stalin persisted in his moderate stance. "The government must not include the exchanged ones," he said, referring to the communists who had been sent to the Soviet Union in exchange for prisoners or sentimental objects. "They will be looked upon as people who depend on Moscow. At home it is a different matter. Let the people elect them."

He agreed that Hungary should have an armed force and offered to release prisoners of war who declared themselves ready to take up arms against
the Germans. He also gave his blessing to agrarian reform. "The estates of
the great landlords have to be given to the peasants." Interestingly though
he advocated this measure not for its inherent merit but on the argument
that Hungary could not be allowed to continue without land reform when
all the surrounding countries had adopted it.[4] On the whole his position, the
only one that really counted, was bewilderingly lenient. He had already
imposed a communist government in Poland, Hitler's first wartime victim;
he would in short order do the same in Rumania and Bulgaria and his agenda
already included the truncation of Czechoslovakia by annexing its eastern
extremity, Ruthenia. In Hungary, still at war with the allies, he wanted a
broad multiparty government and evinced a sensitive concern for its legitimacy; he made no territorial demands for the benefit of the USSR or the
border states. He did not offer an explanation for his forbearance. Personal
sympathy for Hungary's "Moscow communists" certainly didn't explain it
as Stalin had no personal sympathies. Furthermore, all those men, except
for Imre Nagy, were Jewish, and Stalin was at best ambivalent about Jews.
Later developments in Hungarian-soviet relations would shed some light
on this riddle.

Gerö and Nagy, together with Horthy's generals, left Moscow on
December 7 and arrived in Debrecen, a city in east-central Hungary, on
December 12. Between December 15 and 19 elections were held in about
fifty large and medium-size towns for a Provisional National Assembly.
The assembly met on December 21 in the oratory of the Reformed (Calvinist) College. It had 230 members, drawn heavily from lower-middle-
class ranks. Fifty-six percent were industrial workers, journeymen, clerks,
tradesmen, and poor peasants; 22 percent were intellectuals, 16 percent
prosperous agrarians, while 6 percent were priests, military officers, or
men of indeterminate position. All but 5 percent had some party affiliation.[5]

Communists formed the largest single group in the assembly: a total
of 90 delegates, or 39.1 percent. There is no evidence however that the
fledgling party had exerted pressure during the elections or tried to exclude unfriendly voters or delegates. The large number of votes it received can only be explained by the presence of the Red Army and the
psychological effect it created. Gerö, writing to Mátyás Rákosi in Moscow
on December 28, reported (on the basis of information received from
members of the assembly) a total Communist Party membership of 3,053
but added: "This is only a fraction of our real strength. Part of the coun-

try is not represented [in the present membership] at all. And since we are under instructions to be more liberal in our admissions policy, we may within a few weeks have 10,000 or 15,000 members. Our prestige grows by leaps and bounds. We are the chief movers of political life."[6]

The National Assembly's first appeal was to the Hungarian people for whose destiny it made itself responsible. On the very day of its formation it issued an emotional manifesto. "For the salvation of our nation bleeding from a thousand wounds and finding itself in mortal danger, an extraordinary effort is necessary. . . . [We] cannot be allowed to sink into a grave to whose very edge we have been dragged by our accursed alliance with Germany. We do not want to witness in consequence of a war at the side of the German conquerors . . . that the work of generations be destroyed and the thousand-year-old structure of Hungary should collapse." (The reference to the thousand-year-old structure was unfortunate, an insult to the neighbor states who had stripped away large chunks of that structure after World War I, but the drafters of the manifesto had not yet caught the spirit of the times.) The text concluded: "We must break with Hitlerite Germany! . . . We cannot passively watch as the Hungarian fatherland is liberated by the Russian army from the German yoke. We truly deserve freedom and independence only if we actively, with all our power, partake of our liberation. Let the Hungarian people rise to its feet in a sacred war of liberation against the German aggressors!"[7]

On the next day, December 22, the assembly elected a Provisional National Government. Of the twelve portfolios the communists held three, the Smallholders and the Social Democrats two each; one minister was a member of the Peasant Party and four had no party affiliation. The premier was a colonel general, Béla Dálnoky-Miklós; the foreign minister, destined to have the longest tenure, was János Gyöngyösi, a Smallholder, a former bookseller in the town of Békéscsaba.

The second document issuing from the assembly, on December 22, was a declaration directed to the allies; it delineated a far-reaching program for the regeneration of Hungary. It too pledged "to break once and for all with the German oppressors who for centuries have subjugated the Hungarian people." It identified as its most pressing task: "To conclude an armistice with the Soviet Union and with all those freedom-loving nations with whom Hungary is in a state of war." It promised, in its final paragraph, "to recompense the damages the Hungarians had inflicted on the Soviet Union and on the neighbor states through the war waged against them."[8]

The actual request for an armistice was made by Foreign Minister Gyöngyösi to the soviet plenipotentiary in Hungary (a member of the political council of the Second Ukrainian Front), Georgy Pushkin. The request was accompanied by a promise that Hungary would participate in the struggle against Nazism "and thereby atone for the crimes of the previous government."[9]

(Moscow later argued that by merely stating its intention to wage war against Germany but not immediately declaring it, the Hungarian government cast doubt on its sincerity—but it is unlikely that this fact had a vitiating effect on the armistice terms the government received.)

On Christmas Day, in Moscow, Dekanozov informed U. S. Ambassador Averell Harriman of Hungary's request for an armistice.[10] The next day Molotov invited Harriman, and his British colleague Arthur Balfour, for a discussion of the terms.[11] In asking Washington for instructions, Harriman reminded his superiors of the unhappy experience of the western powers in Rumania and Bulgaria, where the three-power Allied Control Commissions (ACCs) were under effective soviet domination, and added: "I am fearful that unless we register our demands in regard to the status of our representatives in Hungary before we agree to the Hungarian armistice, not only will our representatives receive the same treatment in Hungary, but also the Soviets will interpret our failure to insist on our demands as acquiescence in the status of our representatives on the Control Commissions in Bulgaria and Rumania."[12]

Moscow anticipated western misgivings by prevailing on the Debrecen government to change the composition of the delegation it was about to dispatch to Moscow to receive the armistice terms. Imre Nagy, a communist, was replaced by István Balogh, a Catholic clergyman who would soon become the butt of communist antichurch propaganda. The other two members of the delegation were Foreign Minister Gyöngyösi, a Smallholder, and Defense Minister János Vörös, an army general with no party affiliation.

In the meantime, in Moscow, Molotov on the one hand and Harriman and Balfour on the other were engaged in intense negotiations regarding the precise terms of the armistice agreement, The most contentious points were the amount Hungary was to pay in reparations and the status of the British and American representatives on the ACC. As to the first point, the western powers were of course aware that every dollar the strapped Hungarian treasury paid out for reparations would diminish its capacity to indemnify western interests for the losses they suffered during the war due to damage and confiscations and rehabilitate enterprises in which western firms had vital interests. Molotov originally demanded $400 million in reparations. Finland and Rumania had been assessed $300 million each, but they paid only to the Soviet Union, whereas Hungary also owed reparations to Czechoslovakia and Yugoslavia in the estimated amount of $100 million. At the insistence of the western ambassadors the total amount was eventually reduced to $300 million, two-thirds of which was to go to the USSR and one-third to the smaller states. The period of payment was increased from five to six years.

Another contested point was the price level at which deliveries would be credited. Harriman and Balfour wanted to use current world prices, which, due to wartime scarcities, were absurdly inflated. Molotov proposed crediting deliveries at the stable 1938 prices. In the end he agreed to an increase of that price level by 15 percent on capital machinery and 10 percent on food and finished products. (Unfortunately no provision was entered for distinguishing between reparations and war booty, an omission that would cost Hungary dearly.)

Harriman, though he officially complained that the reparations burden was too heavy, had private thoughts to the contrary on the matter. He wrote to Secretary of State Edward Stettinius on December 31, 1944:

> I must confess that I have some sympathy for the Soviet view that 50 million dollars a year of goods as reparations payments from Hungary over a 6-year period is not in fact excessive, and I also feel that there is real value for us in having the claims of the Soviet Union for reparations fixed at this time as otherwise we might have serious difficulties in the future should their appetite grow. On the other hand it seems clear that the manner in which reparations are completed, the character of goods demanded, and the value placed on them, are all matters which would vitally affect the recovery and stability of the economy of Hungary and Central Europe. Whoever controls reparation deliveries could practically control Hungarian economy and exercise an important economic influence in other directions. The Soviet Government's position that only those countries receiving reparations should be involved in the way in which reparations are collected does not seem reasonable. The British and we have an equal interest in the economic stability of Europe, even though neither of us are demanding reparations from Hungary.[13]

On the question of the authority of the ACC, one of great sensitivity to the western powers, Molotov agreed to some cosmetic changes. According to his original formula, the Provisional Government could not take actions of a certain nature without permission of the "Allied (Soviet) High Command."[14] This formulation was changed to "Allied Control Commission." Nevertheless for the duration of the war the government had to seek clearance only from the soviet delegation, which would in turn "inform" the western allies. After the end of the war the entire ACC would have to be consulted. Molotov cited security considerations for the soviet demand that western members of the ACC seek permission from the soviet chairman for leaving or entering the country. The chairman would also have the right to determine the number of officials the British and American delegations were allowed to have. This stricture too would terminate after the war.[15]

The allies had agreed, at different times, that before finalizing the armistice terms they would consult the Czechoslovak and Yugoslav governments.[16] The Czechoslovak ambassador in Moscow left a memo to this effect with the foreign ministry and gave a copy to Harriman. It requested among other things that prior enactments providing for the transfer to Hungary of Czechoslovak lands be declared null and void and that the date of the war existing between Hungary and Czechoslovakia be pushed back to October 7, 1938 (the day when bilateral negotiations for the transfer of Slovak territory commenced).[17] The Yugoslavs demanded only that Hungary renounce for all time territorial claims on their country and that the borders revert to their 1940 status.[18] Some of these demands found their way into two separate articles in the armistice agreement. Article 2 provided that: "Hungary has accepted the obligations to evacuate all Hungarian troops and officials from the territory of Czechoslovakia, Yugoslavia and Rumania occupied by it within the limits of the frontier of Hungary existing on December 31, 1937." Article 19 read: "The Allied Governments consider the Vienna arbitration award of November 2, 1938 [which provided for the transfer to southern Slovakia to Hungary] and also the Vienna award of August 24, 1940 [which detached northern Transylvania from Rumania] as null and void."[19]

In general though the western allies found that they had very limited leverage in negotiating with the soviets about East Europe. When, in October 1943, Churchill made his controversial "deal" with Stalin regarding the division of spheres of influence (providing for a 90 percent British influence in Greece, a 90 percent soviet influence in Rumania, and a 50-50 percent arrangement for Hungary and Yugoslavia) he had offered not a wish-list but what seemed at the time a sober assessment of military realities. Still hoping for an allied front in the Balkans, Churchill saw a chance for a more or less equal allocation of power in East Europe. Since then the Balkan front had become a chimera, the soviets had already conquered half of Hungary, and nothing could prevent them from occupying the rest. (British Foreign Secretary, Anthony Eden, in consultation with Molotov, had already conceded a 75 percent share of soviet influence in Hungary.[21]) The military, and hence political, advantage had shifted heavily to the soviet side. This fact greatly limited western ability to gain points through negotiations.

The Hungarian delegation had arrived in Moscow on January 1 but had to wait until the great powers had ironed out their differences. Gyöngyösi received the text of the armistice agreement on January 18. He was granted but one day to study it and to formulate his objections. In addition to the articles already mentioned, others provided for the obligation to disarm German soldiery on Hungarian territory and to turn over all seized weaponry and stocks to the Red Army. German civilians on Hungarian soil were to be interned. Hungary agreed to set up eight infantry divisions to be deployed against German forces; these divisions would have to be disbanded after the war. The Provisional Government would have to guarantee complete freedom of movement to soviet and other allied personnel and place all necessary means of transportation at their disposal at its own cost. Allied prisoners of war would at once be freed and allowed to return home. By Article 5 the Provisional Government obligated itself to "ensure that all displaced persons or refugees within its territory, including Jews and stateless persons, are accorded at least the same measure of protection and security as its own nationals."

Gyöngyösi offered only two comments on the draft. One concerned the fate of Hungarian prisoners of war; the other was the request that reparations obligations be stretched out over a ten-year period instead of the six years specified in the draft. Molotov left the first question unanswered and rejected the other.

On January 20 Gyöngyösi put his signature to the armistice agreement, as did representatives of the three great powers. The Provisional Government, acting in the name of the entire nation, was out of the war with the Grand Alliance but was at war with Germany and by extension with the Hungary still ruled by the Szálasi government. The country had not regained its sovereignty by leaving the war because an Allied Control Commission oversaw all its internal and external affairs and the implementation of the armistice terms. The all-powerful chairman of the commission was soviet Marshal Klementi Voroshilov, with Georgy Pushkin acting as his liaison with the Hungarian government.

Budapest was still under siege and the Buda hills were being bitterly contested but the eastern half of the country could breathe free again.

The armistice agreement, to be sure, was a temporary expedient with little relation to the country's long-term future. Hungary's quarrels were not with the Soviet Union or with Britain and the United States but with its neighbors to the north, east, and south. These quarrels dated back centuries and today, half a century after the armistice, are as alive as they were then.

The archives of the Hungarian foreign ministry contain an undated memorandum, probably prepared for presentation at the peace conference, which although tendentious, presents a generally accurate picture of Hungary's place and destiny in East Europe. Over that area, as large as France and Spain combined, the memo explains, dwell 12 major ethnic groups, side by side or mixed together. Each of these groups numbers half a million or more, and there are eight or ten nationalities of lesser size. The geographic core of the region is the Carpathian basin, a closed-in area of about 350,000 square kilometers; this is where north and south, east and west meet. For centuries it had been the area of contact, as well as strife, between Romans and Byzantines, Christians and Moslems, Germans and Slavs. From the ninth century onward it was largely the Polish and Hungarian settlements that had provided the area with stability and a balance between east and west.

Of these two nations (both ruling over a number of lesser nationalities) Hungary was indisputably in the better position. It was well-suited for defense, had a varied topography, and its regions were well-connected. It proved one of Europe's most durable states, its borders barely changing over the centuries. Poland by contrast had no clearly defined frontiers or convenient means of internal communications. Constantly changing its shape, its political organization remained loose and unstable.

In the sixteenth century Hungary, or the central portion of it, fell to the Turks; at the same time, because of the death in battle of its childless young king, it passed under Habsburg domination. In the 1700s Poland on its part succumbed completely, partitioned between Russia, Prussia, and Austria. Hungary had by that time been liberated from Turkish rule by a long and devastating war that left the area over which it was waged largely depopulated. Settlers poured in from neighboring countries. Most came from Serbia and Rumania, two countries that remained under Turkish rule. The settlers were lured by the open, fertile Hungarian plains. The thoroughly mixed ethnic composition of the Carpathian basin dated from this time.

Ethnic confusion was of course characteristic of Europe as a whole; the difference was that in countries such as Spain, France, Italy, Germany and even England the diverse language groups were so jostled together that in time the difference among them disappeared. There was a great deal of jostling in the Carpathian basin too, but the Hungarian element remained dominant. Then, nearly two centuries after the Turkish occupation and the attendant depopulation, the country experienced another wave of migrations; during the eighteenth and nineteenth centuries so many new settlers arrived that by the time of the outbreak of the Great War the grinding-down process had not been completed.

In the nineteenth century, with nationalism at its height, the various ethnic groups, far from accepting assimilation, sought to keep themselves apart and reaffirm their ethnicity. The area of the Magyar settlement shrunk; it was concentrated in the Great Plains in the center of the Carpathian basin, in Transdanubia, on the Little Plains just north of the Danube, and in the southeast corner of Transylvania. Until that time the Rumanian population had been confined largely to the trans-Carpathian province of Moldavia and to the high mountains of the eastern Carpathians; in the nineteenth century it rapidly extended into Transylvania. Meanwhile Serbs and Croats pushed northward into the Great Plains and mingled with Slovaks and Ruthenes coming from the north.[21]

Another document preserved in the same file and preparatory for the peace conference points out that despite the ethnic mosaic all these migrations produced, it was still possible after the Great War to draw borders that effectively separated one language group from another. But the paramount consideration at the Paris peace conference was not to draw ethnically fair borders but to reward the victors at the expense of the defeated states. One observer of the peace process, Harold Nicolson of Great Britain, later wrote that the great failure of the conference was that, instead of considering the total picture in East Europe, it dealt with the claims of individual states separately.[22] This held particularly true for Hungary. Many border disputes could have been resolved by minor modifications. Instead, the heterogeneous Habsburg Empire was broken up, not into homogeneous but into smaller heterogeneous states, sharpening instead of lessening ethnic tensions.

The area left to Hungary after the peace settlement (Trianon Hungary as it came to be called in contrast to the "historic kingdom" or "the lands of the Crown of St. Stephen") lay at the very bottom of the East Central European basin and was densely populated—in Europe only the Netherlands, England, Germany, Czechoslovakia, and Switzerland had greater population density and those were industrialized and had ample natural resources. As part of the Dual Monarchy before 1918, Hungary had easy access to finished goods and neglected industrialization. After the war the sources of manufactured articles had passed to hostile neighbor states and Hungary tried to make up decades of neglect, but with little success. By the 1930s fully half of the country's population still lived on the land, while mining and industry employed 2.8 million people of a total population of 8,698,000. Yet agricultural products, even at the end of the 1930s, accounted for only 37.2 percent of the gross national product, mining and industry accounting for another third. As for exports, agricultural and forestry products made up 60 percent of the total, industry 37 percent, mining and smelting only 3 percent.

The failure of industrialization was due largely to a lack of raw materials and capital. The country, reduced to one-third of its former size, had only one iron mine and a very low supply of nonferrous metals. Raw materials for the textile and leather industries had to be imported and Hungary had no salt at all. Even agriculture suffered from the fact that most of the "rested" soil had been transferred to neighbor states, especially Czechoslovakia and Yugoslavia. It was in these areas, contiguous to the native land, that Hungarian minorities were most densely settled, forming an absolute majority of the population. Their resentment at being separated from the fatherland was keen and unabating.

The drastic truncation of Hungary's land area, with all the dislocation of people and resources it entailed, remained the central problem even after World War II. Between the two wars the loss of territory was generally looked on as temporary, the result of grievous mistakes by the peacemakers who were blind to the realities of the region: thus adjustment to the changes was reluctant and tentative. Very few hoped, to be sure, that the "historic kingdom" could be restored in its entirety, but the redemption of as much of it as could be achieved without paying a prohibitive price remained a solemn patriotic duty. How this passionately held purpose gradually brought Hungary into the Axis orbit despite the resistance of many of its leading politicians and how that policy culminated in Hungary's participation in a losing war has been dealt with in many thoughtful studies and a recapitulation of even the principal points would be redundant. A curious anomaly must be pointed out however. When Hitler launched his own revisionist campaign in 1938 he argued for it not on historic but on ethnic grounds. By this he implicitly adopted Woodrow Wilson's program of national self-determination. It was, to be sure, a tactical expedient, but it set the tone for all the other revisionist campaigns.

When Hitler first reached beyond the borders of Germany proper, into Austria and the Sudeten areas of Czechoslovakia, he posed as the champion of ethnic Germans under foreign rule. In an attempt to apply maximum pressure on Czechoslovakia, he also took up the cause of ethnic Poles and Hungarians who were troublesome minorities in the republic. Thus he foisted on Hungary a program that aspired only to territories with a Hungarian population: it was precisely this that nationalistic elements found unacceptable. One of their spokesmen put it bluntly: "For us[Hungarians], who are unwilling to relinquish our role in the Danubian basin and who are both destined and able to rule foreign peoples . . . the adoption of the ethnic principle would be suicidal."[23]

This however was rhetoric. Hitler at the time was the sole arbiter of territorial revision and he wanted a finely balanced mix in the Danubian region so he could play off one nation against another. The western powers, by the infamous Munich Agreement, in effect surrendered their interests in East Europe. An appendix to that agreement provided that if Hungary and Slovakia did not settle their dispute over the ethnic areas within three months, it was to be arbitrated by the four Munich powers. But when such arbitration became timely, Britain and France politely declined to participate. On November 2, 1938 the foreign ministers of Germany and Italy, meeting at Vienna, drew a line on the map, awarding the southern strip of Slovakia to Hungary. The stricture of the Trianon treaty was broken. Four and a half months later, as Hitler made ready to destroy what was left of Czechoslovakia, he invited Hungary to occupy still another of its provinces, Ruthenia.

Hungary's revisionists now turned their eyes to Transylvania. Some two million of their kin dwelt in this large and richly endowed province; the southeastern corner of it was home to a solid ethnic bloc of Szeklers who had lived there for centuries and spoke Hungarian. It was this group, located so far from the country's core area, that made the drawing of a borderline so problematic. The Paris peace settlements awarded the entire province to Rumania. But the Paris peace settlements had fallen on hard times. In the summer of 1940 Hitler was actively considering a campaign against Russia and he wanted no unsettled nationality problems on his wings. In August he yielded to Hungarian demands for the restoration of at least part of Transylvania. The result was an awkwardly arching protuberance from Trianon Hungary to the Carpathian corner inhabited by the Szeklers. Seven months later, as Yugoslavia fell victim to Hitler's Balkan reordering, another lost province of Hungary, the Bačka, enclosed by the lower flow of the Danube and Tisza rivers, was joined to Hungary as a reward for its entering the campaign.

Operation Barbarossa, as Hitler's campaign against the Soviet Union was codenamed, opened in June 1941. There followed a scramble by lesser states to earn merits for future territorial redistribution by joining the war. Slovakia, by now an "independent" country, declared war on the Soviet Union, as did Rumania. Hungary delayed for five days. When it joined it found itself in alliance with two states from whom it had recently taken land and who were determined to regain it.

An initial force of two army corps was sent to the eastern front in the expectation that, as had been the case in Hitler's previous campaigns, German victory would be swift and decisive. When the battle bogged down in the depths of Russia and there was no prospect of an early end, Hungary withdrew its modest force and left only some occupation troops on Russian soil. The war was unpopular in the country. Hungary had no quarrel with the Soviet Union and no claim to any of its territories. The realization dawned that Hungary was fighting Germany's war; this meant that it had lost at least part of its independence.

Hitler at first did not object to the Hungarian withdrawal from the front, but when victory eluded him, he pressed Hungary, as he pressed his other allies as well, to send new armies to the eastern front. It would have been hard to resist the pressure in any case, but there was also the danger that failure to comply with the German demand would lead to a renewed loss of the recovered provinces. In the end one of the country's two armies, the Second, was sent to Russia with a manpower of 300,000. Eight months later a horrified nation watched as that army was shattered on the Don in the aftermath of the great Russian counteroffensive at Stalingrad.

From now on it was the allies' war. Strategic direction passed from Hitler's hands to those of Stalin, Churchill, and Roosevelt. More than ever the small states became pawns in the gigantic struggle of the great powers. As the Red Army rolled forward irresistibly, Hitler's allies deserted him one by one. The tattered remnants of the Hungarian army fought on. The entire nation that had, unenthusiastically but still loyally, supported the German war effort and now stood on the brink of defeat, braced itself for a new harsh and punitive peace.

Chapter 3

Groping toward Democracy

Among the defeated states Hungary bore the heaviest burden of war guilt, because of its belated appeal for an armistice and because of the continued belligerence of the fascist government in the west. When the leaders of the allied powers met in conference at Yalta two weeks after the conclusion of the armistice, there was ample evidence of their lack of sympathy. Churchill and Roosevelt were assertive enough in their attempts to save Poland, Rumania, and Bulgaria from falling under soviet hegemony, but they showed little interest in Hungary's fate. A leading Hungarian diplomat bitterly remarked after the conference: "the western allies of the USSR tacitly abandoned us to the soviets."[1]

What some called abandonment, the communists called liberation. It became the operative phrase and any alternative was an insult to soviet sensibilities. A town or city never fell, it was liberated. The entire country was being liberated, not only from the Germans but also from its selfish and exploitative ruling classes. When Mátyás Rákosi arrived back in Hungary with members of the armistice delegation on February 1, 1945, he announced a figurative gift to the laboring people of Hungary: "The country is yours, you build it for yourself."

The country indeed needed rebuilding on an inestimable scale. The ravages of war and outright plunder had left it destitute. The greater part of the equipment and machinery of 415 factories had been dismantled and carted away by the Germans and their Arrowcross allies. The transportation system was paralyzed by the loss of rolling stock: 37,160 railroad cars, 26,000 automobiles, 5,600 trucks, and 270 water-borne vehicles, as well as 8,000 freight cars laden with food and dry goods, had been shipped to the Reich. The entire equipment of 25 major hospitals and university clinics had been taken away. On orders of the Arrowcross president of the National Bank the entire gold supply, the numbered and cut banknotes, all negotiable instruments, and even the equipment of the mint had been transferred to Germany. The loss in livestock, slaughtered or driven away, was staggering.

By conservative estimates the total value of goods lost or stolen amounted to some $13 billion in 1938 prices. Most of the displaced assets were in the American occupation zone in Germany, and in Austria, but there were sizable quantities in the other three zones as well. The government made it clear early that the return of these goods was a precondition to any substantial recovery. And without such recovery Hungary could not satisfy the many obligations it had undertaken in the armistice agreement; these included, apart from reparations payments, the compensation of citizens and enterprises of United Nations countries for the losses they had suffered in consequence of Hungary's partnership with Germany. With no reserves to draw on, reparations payments had to be made from current production at a time when the industrial plant had been reduced to less than 20 percent of its prewar capacity.

In these conditions Hungary needed international aid and goodwill on a grand scale and there was little of it at the moment. The British and French remembered Hungary as a despoiler of Czechoslovakia in the wake of the Munich Agreement and an aggressor against Yugoslavia in its gravest hour. The United States evinced occasional interest in conditions within Hungary but this was more an attempt to limit soviet influence in the country than a genuine display of goodwill. Meanwhile soviet soldiery continued its exactions and the working people suffered as much as their capitalist "exploiters" did. The neighboring states were hostile and systematically persecuted their Hungarian minorities, sometimes as official policy, sometimes by allowing free hand to illegal bands. The prime task of Hungarian diplomacy was to build bridges to the outside world and to project the image of a country chastened by defeat, ready to embark on a new, progressive course.

Yet the foreign policy establishment itself was decimated. Many of its former officials were out of the country, others were in hiding. Still others, who had served the Nazis, were politically unacceptable. At the time of the formation of the Provisional Government the foreign ministry occupied three rooms in Debrecen's old financial center. János Gyöngyösi had a staff of two clerks to assist him. The prior ministry building in Buda had been bombed out; buried under the ruins were many of its documents as well as its priceless library. A few former officials came together and began to clear away the rubble. But the reputation of many of these diplomats and functionaries was under a cloud and for now the rehabilitation of the ministry as a whole was an impossible task. The coalition parties agreed to screen the personnel roster left over from the Horthy years and decide who was politically fit to be reemployed. At the same time criteria were established for hiring new employees. As party affiliation often counted for more than expertise, some unhappy choices were made.[2]

The status of the diplomats still abroad was unclear; Ferenc Szálasi had kept most of the Horthy appointees and now the new government was at a loss how to proceed with their accreditation. In a legal sense, there was no diplomatic service at this point. Not until August 1945 did the executive division of the ministry announce its plan to set up altogether eight legations: in Moscow, Washington, London, Paris, Prague, Belgrade, Bucharest, and Vienna. There was as yet no provision for the establishment of consulates.[3]

An internal memorandum accompanying the recommendation wryly observed that there were very few people suited for the position of minister or ambassador. Those in place, although almost without exception of the old school, would be hard to replace. "For this reason we respectfully inquire whether we might propose officials who had been given minor punishment during the screening process but who are politically not seriously objectionable."[4]

In point of fact what passed for foreign policy was largely limited to contacts with the Allied Control Commission (ACC). But in February, and even March, of 1945 that body was not yet complete. According to the interallied agreement it was to begin functioning as soon as the armistice agreement became effective. However, on January 20 the designated chairman, Marshal Klementi Voroshilov, was still in Moscow (he was represented at Debrecen by a Major General I. I. Levushkin) and members of the western delegations had not arrived. The man with the greatest political weight in the temporary capital was the de facto soviet envoy, Georgy Pushkin.

Budapest fell in early February after extremely heavy fighting. A good part of the country still remained under Arrowcross rule, but to the minds of most Hungarians Budapest was the heart and the functional nerve center of the country and its possession was the ultimate test of a government's legitimacy. When Voroshilov arrived in Debrecen in early February, he found on his desk a letter from the president of the National Assembly, oozing with phrases of humble gratitude and ending with the request that the assembly soon be allowed to transfer its seat "to the country's liberated heart [where] it can securely undertake and complete its great task and lay the foundation for a new, independent, democratic state organization, freed from foreign influences imposed on us over many decades, relieved of the burden of the past, its sins and mistakes."[5]

Early in March, the premier, Béla Dálnoky-Miklós, formally raised the question of moving the government to Budapest, arguing that the move was of pressing importance, "politically and administratively." A Russian general present at the interview opposed the request, saying the front was still too close.[6] Three weeks later Voroshilov vetoed a similar request, claiming that the city was in bad shape, the buildings were in an advanced state of disrepair, food was not arriving, and factories had closed down for lack of fuel.[7]

The British and American delegations had not yet arrived and Voroshilov took it upon himself to lay down the law to a cabinet committee consisting of the premier, the foreign minister, and the defense minister. His principal instruction was that all communications with nonsoviet ACC personnel had to be made through him. He then cited two articles of the armistice agreement that were of particular interest to his government: first, that all assets Hungarian armies had removed from territories they had occupied had to be restored; and second, that the government had to furnish a list of war criminals. This drew from Dálnoky-Miklós the question of whom the allies regarded as a war criminal. Voroshilov replied that the question was still under consideration. He then asked whether the government was at last ready to man the various administrative functions at lower levels, specifically those charged with carrying out the armistice terms. The premier replied in the affirmative but added that the demands made in connection with armistice provisions were often so excessive that the officials were unable to satisfy them.[8]

A pattern was emerging: the chief benefit the soviets sought to gain from their occupation was the ability to lay their hands on whatever movable asset was still available and claim it under the rubric of war booty, or as reparation. And most confiscations were made without any legal ground at all. During the spring and summer barely a week passed without the foreign ministry or the premier's office complaining to Voroshilov, or to Pushkin, over the liberties soviet authorities were taking. Truckloads of food sent to Budapest from the provinces were apprehended by soviet military units and driven away.[9] Hundreds of horses were conscripted for traction then kept or turned loose.[10] Timber shipments intended for the restoration of rail lines were seized by soviet soldiery and confiscated.[11] The list was endless and the soviet military made no distinction between private and government property. Banknotes and foreign currency taken to Austria by the Arrowcross government had been seized in Vienna by soviet forces; their total value was $21 million.[12]

At first Voroshilov and Pushkin received these protests with a show of concern and promised to investigate them, but in time their patience wore thin. At one point Voroshilov brusquely requested the premier to instruct the foreign minister to cease sending his "notes" to the ACC; "they are largely unfounded and refer to questions which do not belong into the purview of the foreign ministry." In any case, he added, there was no substantiation in a single instance to support the allegations.[13]

Gyöngyösi did not venture to answer that. But in an internal memo he noted that his latest protest had been occasioned by a concrete incident that was only one of countless others. "It happens daily that local Red Army commanders request from provincial authorities deliveries that are not entered into the central register and hence are not part of the accounting," he wrote. As for the rest: "The foreign ministry cannot limit its functions to foreign affairs. It is its duty to protect the country and its citizens in every matter that has foreign implications. This duty it will not abdicate."[14]

By March the personnel of the soviet contingent assigned to the ACC reached 750, by far the largest of the three powers. Two-thirds of the staff, as a soviet official confided to a Social Democrat, consisted of technicians and specialists. This was explained by the fact that the soviet government intended to liquidate a good part of Hungary's reparations obligation by carting away industrial machinery, a process requiring a great deal of technical expertise.

There were two million soviet soldiers in Hungary and after years of privation their rapaciousness was hard to control. This too was a foreign policy item, albeit not one that could be settled through diplomatic means. In general, there was no bipartisan or multipartisan consensus as to what the country's foreign policy should be; a deep ideological cleft separated the political left from the right. Communists, standing on a platform of international workers' solidarity, professed themselves to be realistic about the conduct of soviet occupying forces. In their eyes suffering occasional illegalities was a small price to pay for the country's liberation. Some 400,000 soviet soldiers had died in battles against German and Hungarian fascists on Hungarian soil. The soviet people had bravely shouldered this loss, recognizing their duty to bring freedom to oppressed people everywhere.

The "reaction" on its part based its judgment on concrete realities. Being freed from German occupation and being liberated were not necessarily one and the same thing. This was what former premier Miklós Kállay had referred to in his emotional assessment of Hungary's position. The Germans had inflicted their own bloodletting but it had been confined to the Jewish population. That was a "sacrifice" the rest of the nation had accepted with equanimity. Pilfering abandoned Jewish wealth had been a virtual industry in Hungary and vacated Jewish homes accommodated the streams of refugees pouring in from Rumania. However Soviet outrages hit everybody equally hard. The illegal robbery was even worse than the authorized confiscations. It took audacious rhetoric to term all this liberation.

The forever contentious territorial issue also would not go away. The reaction took the position that military defeat did not invalidate Hungary's claims on its neighbors. The wrongs inflicted on the country at Trianon did not become more defensible because Hungary had lost a second war. It was a matter of national honor to demand back the territories adjacent to Hungary in which the bulk of the population spoke Hungarian; or at least these people must be granted full civil and political rights. The communists denounced this position as a residue of an old and discredited past. In their view once Marxist principles gained ascendancy, quarrels over strips of land would become irrelevant.

These ideological differences had of course broader implications. Communists saw Hungary's salvation in the closest possible ties with the Soviet Union and they even regarded the presence of western representatives as a nuisance and an encouragement to the reaction. The political right on its part looked on soviet occupation as a dreadful imposition that stifled the country's normal development and mercilessly plundered what little was left of the nation's wealth.

Meanwhile those who kept a close watch on the policies of the Hungarian Communist Party (MKP) as a clue to soviet intentions could read some encouraging signs. The extremist elements who had argued for the immediate establishment of a soviet republic to be maintained by terror had been silenced. The lessons of history weighed heavily and the leadership knew that such a republic would end up as the Bolshevik regime of 1919 did, though the presence of the Red Army might give it a longer lease on life.

Thus the leading party cadre depended on propaganda rather than terror. That this propaganda, at any rate in the early months, lost much of its credibility due to the behavior of soviet soldiery was something neither the Hungarian communists nor the soviet command could do much about. Deportations of civilians continued unabated. Having a German name was often reason enough for a person to be taken to parts unknown. Zoltán Vas wrote in one of his letters: "They are taking away

the Germans, probably to the Soviet Union. Let them. But they are taking away Hungarians with German names as well. We barely succeeded
in rescuing a few excellent party cadres." Ten days later: "Not only do
we suffer, the cause suffers as well."[15]

When Rákosi arrived in Debrecen he wrote to the party cadres in
Budapest: "Great tasks await us, but we know the road and we know the
goal. . . . Let us with all our strength enhance party discipline. Let us provide an example [and] . . . sacrifice in every respect. . . . Let us keep corrup-
tion away from our ranks, let us mercilessly remove all those guilty of it."[16]

But the enormity of the task was clear to him and he unburdened himself in a series of letters to the former head of the disbanded Communist
International (Comintern), Georgy Dimitrov. The party in Hungary, he confessed, was shapeless and directionless. "There is much factionalism and a
lack of experience in working with the masses. The overwhelming majority
of the rank-and-file consists of sentimental communists who will naturally
have to be remolded. There is no firm party core to speak of."[17]

He too lamented the outrages of the liberators. "The excesses of the Red
Army become the party's liability. . . . Cases of mass rape of women and of
robbery are repeated with the liberation of each region, as recently in
Budapest too. Raids [in the streets] are being routinely conducted, in the
course of which workers, some of them good party members, are taken to
prisoner of war camps where they disappear."[18]

The deplorable conduct of the occupiers reflected not only on the Communist Party but on the entire government, which in the eyes of most Hungarians was feckless and indifferent. Yet it was just now that it needed all
the prestige it could command because, even as the war was still being
fought, it decided to undertake a measure that, in view of Hungary's feudalistic past, was probably the most radical ever introduced: a thorough land
reform. Back in 1918, at the close of the Great War, a progressive nobleman
had begun a modest redistribution of land, but the attempt was overtaken by
events and after Horthy's seizure of power was abandoned altogether. It
was precisely the contrast between the great landed wealth of the aristocrats
and the grinding poverty of the agricultural proletariat and the small peasantry that had branded Hungary in the eyes of the outside world a land of
medieval backwardness.

The first proposal for a general land reform had a mixed reaction, among
the political parties as well as western representatives. The communists would
have been expected to be in the vanguard of such a program, but they were
at first halfhearted and hesitant. Probably the apprehension of being identified too early as a party sponsoring radical measures kept Rákosi and his

colleagues from being too vocal on the issue. But by mid-May 1945 the party had reason to regret its caution. An internal report that recounted the National Assembly's major exactments was in fact mildly critical:

> Admittedly [in this question] we were not decisive enough. Our target date was October 1. Such a delay could have been fatal. But we recovered and decided to carry out the reform in six weeks. [This quick] implementation prevented the reaction from organizing resistance. By the time it recovered, it was too late.[19]

The real moving force behind the measure was Marshal Voroshilov and he had primarily the exigencies of war in mind. Perhaps he harked back to the Russia of 1917 when the Bolsheviks carried out successful propaganda in the armed forces, hinting at land distribution in the hinterland to encourage the defection of poor peasants who feared that if they arrived home late all the land would have been parceled out. He now wanted to provide an incentive for the soldiers fighting in Szálasi's armies to desert; he urged the government to announce a land reform without delay. In the countryside Red Army officers helped form committees of poor peasants to agitate for the measure. The bourgeois parties, including the Smallholders, into whose ranks most of the anticommunist elements had by now entered, wanted to delay the reform until the plans had been worked out in greater detail, and this was a position to which the western powers gave full support. When the American representative, Arthur Schoenfeld, and the British representative, Alvary Gascoigne, protested to Gyöngyösi against the breathless tempo of the redistribution, their professed reason was concern over the property rights of foreign citizens.[20] But it was clear to all concerned that their disapproval went deeper.

The hasty division of the landed estates while the war was still in progress confirmed fears that occupation by the Red Army would lead to much broader sovietization. Gascoigne in his report to the Foreign Office declared the land reform not only objectionable but downright illegal, because there was no special law to regulate it.[21] The measure also had an aspect of reverse discrimination. Large landowners were allowed to keep only 100 hectares of agricultural land while small landowners could keep, or received, twice that amount.[22]

The western powers perceived a Jacobinist tendency in all this and it was not so much the economic as the political implications that worried them. The principle of confiscation had received official sanction and it was only a matter of time before it would lead to wholesale nationalizations.

Gyöngyösi was entrusted with informing Voroshilov of the decision to parcel out the large estates. The latter got out of his chair and shook Gyöngyösi's hand. The two men then discussed how the land reform should

be announced to the public. Voroshilov wanted the entire ordinance to be published in the newspapers; Gyöngyösi preferred to give the press only a condensed popular version, then distribute printed leaflets with the full text. Voroshilov promised that the news would reach that part of the country still under Arrowcross rule. He wanted the government to issue a proclamation with promises that landless peasants who deserted from Szálasi's armed forces would be given land while those who fought on would lose the land they possessed.

The interview was friendly and Pushkin, also present, handed Gyöngyösi a letter from the soviet government promising large food deliveries to Budapest and other industrial centers: 15,000 tons of bread, 3,000 tons of meat, and 2,000 tons of sugar. This would be a loan, not a gift, and it would have to be repaid by November 1.[23]

There was political calculation in this seemingly selfless gesture: the capital and industrial cities were citadels of communist strength, yet it was precisely these that were cut off from a regular food supply, in part because of transportation difficulties and in part because of soviet confiscations. The MKP had long suspected that the country was deliberately withholding food from "red" Budapest. Such a suspicion was impossible to confirm or to refute because of the continued ambivalence of the population toward the communists.

The MKP undoubtedly possessed a dynamism and a reformist spirit that invigorated the long-stationary political scene. Many people were willing to reserve judgment about the much-maligned party until it showed its mettle in sorting out the problems that confronted the country. But disillusionment set in early. We already spoke of the effect the presence of a locust army of two million had on the popular mood. At the same time there was also an awareness, especially among the working people, that communists had labored and suffered for decades in an effort to bring a more equitable social order to the country. When the premier, Dálnoky-Miklós, sent Voroshilov an aide memoir in early summer he expressed both sentiments, in somewhat extravagant language. He voiced gratitude "for the goodwill the soviet government has demonstrated in the solution of Hungary's gravest problems." It was virtually without parallel, he wrote, "that a victor state should treat a conquered nation with such magnanimity and farsighted generosity."

However, he then continued without transition in an entirely different vein, furnishing a long catalog of soviet lawlessness. He reminded the marshal that it fell to a destitute country to feed a huge occupying army and that the food supply was consequently exhausted. The industrial plant lay in ruins. What was left was being dismantled by the soviet army and carried away either as war booty or as reparation. The letter ended with this warning: "The faith of precisely that portion of the . . . working class which the Hungarian government most depends on will be undermined."[24]

The inability of the Provisional Government to establish its trustworthiness on the domestic scene was duplicated in the foreign policy field; measures designed to please the soviet occupiers almost invariably incurred the displeasure of the western powers. During the summer the coalition parties held a series of high-level discussions in which they tried to chart a direction acceptable to all. Most of the speakers advocated a middle course, but the communist spokesman, József Révai, expressed himself in unmistakable terms: "Hungary's foreign and internal policy must base itself on the premise that the democratic people's movement [toward socialism] of enormous depth that is a consequence of the war has not come to an end with the peace; on the contrary, this movement will gain strength all over Europe. And we Hungarians must join in it."[25]

At the time this was but one voice among many, a declaration of intention so vaguely worded that it was hard to perceive a concrete program in it. Other opinions were freely voiced. The position of the Smallholder speakers was that while Hungary had to strive for close relations with the Soviet Union, it had to keep the avenues to the west open and clear; this theme would recur regularly until the very moment of the demise of the Smallholders' Party (KGP) as a political force. The Social Democratic Party (SDP) took the middle ground and in endorsing a western orientation concentrated on Great Britain, which had a Labour government. The voices of the other parties were too faint to hear.

In the armistice agreement Hungary undertook to set up eight heavy infantry divisions for participation in the war against Germany; it became quickly obvious that by the time that task was even only partly accomplished the war would be over; still, plans were being developed with the greatest seriousness. As most of the country's manhood had been conscripted the allies had undertaken to release prisoners of war to fill the ranks of the divisions. The problem was not only numbers but political reliability suited for the times.

Within ten days of the signing of the armistice the MKP and the SDP worked out guidelines for organizing a new "democratic" force. The bulk would come from prisoners kept in camps on Hungarian soil; the rest from "farther east," presumably Rumania and the USSR. A bit of involved arithmetic produced an estimate of 70,000 troops. But a force of eight divisions called for at least 80,000. The study therefore envisioned the recruitment of women, up to the age of 40, if physically fit; men would be expected to serve to age 42.

Equipping the force seemed well-nigh impossible. Legally all military hardware was soviet booty. Its release would have to be negotiated with the soviet authorities. Horses, so essential for traction, were so depleted in number that there were not enough for agricultural work. A commissioned officer corps that could be depended on to command a force that was democratic by soviet standards simply didn't exist. If one was to be created it could be done only through promotions and the inclusion of skilled women. Even so, the problem of making an army that had been raised in a reactionary tradition democratic would remain. The two parties therefore recommended a "cultural education department" within the ministry of defense. Its officers would not be involved in military decisions but would be responsible for setting up a disciplinary code and making sure it was followed.

Candidates for the new army would have to fill out a questionnaire relating to their political past. Did they serve in the Horthy army and in what rank? Did they participate in the Szálasi putsch? Most importantly: "After October 15, 1944 [the day of the putsch] did you in pursuance of your oath to the Supreme War Lord [Horthy] to undermine the Arrowcross government imposed on the country . . . exempt yourself from service to it and thus help your country to avoid the horrors of war?"[26]

On March 17, 1945 Rákosi reported to Dimitrov in Moscow that the first division was ready. He continued:

> We have our own people in the most important positions but we have not yet succeeded in planting our feet in the chiefs of staff. The reaction attempts to establish guard regiments, which are supposed to replace the dissolved gendarmerie. They have asked for 10,000 men for this purpose. . . . Many signs point to the fact that the reaction is spoiling for an armed battle. We even have information that they recruit into the army first and foremost fascist noncommisioned officers (NCO); we know of secret instructions not to accept communists or anti-fascists into the army. We of course take the necessary countermeasures.

About the police, the main instrument of communist control, Rákosi had this to say:

> The SDP and other parties are making their voices heard. They grumble about communist monopoly and demand "democratic parity." As most of the leadership positions in the police are in the hands of our comrades, we will have to make concessions. . . . We will of course take care that the key positions remain in our hands. . . . In this question we will not yield."[27]

Rákosi's letters, like all of his other nonpublic pronouncements, have to
be read with the understanding that an intense political bias, verging on
paranoia, informed them. He was prone to ascribe every untoward event,
every failure of a program dear to communists, to the machinations of the
reactionaries. He didn't trust the less-tainted parties or individuals either—
including the Social Democrats, although for decades they had been the
only party openly fighting for the workers' cause. (Of course communists
always cast a jaundiced eye on the SDP, which had been sanctioned by the
Horthy regime on the condition that it would operate as a loyal opposition;
within that parameter it could have a free press and run in parliamentary
elections.[28]) Rákosi wrote to Dimitrov: "So far as our allies are concerned,
the Social Democratic Party, even though it has created a liaison committee
with us in Debrecen and Budapest, has reached a secret understanding with the
Smallholders." This understanding, in Rákosi's judgment, had a history:

> Since 1942 both [parties] had counted on two variants: Either the
> Anglo-Saxons occupy Hungary, or Horthy succeeds in carrying out a
> coup similar to the one which took place in Finland and Rumania. The
> third possibility, that the USSR liberates Hungary, did not figure in
> their plans until the last months and now they are distressed to see that
> the MKP has anticipated them.[29]

Social Democrats on their part heartily reciprocated Rákosi's distrust.
They too had been in hiding during the Nazi era; when they emerged they
resolved to be the main link between the Provisional Government and the
victorious allies. They established within the party a foreign policy and
nationalities bureau with the express purpose of gathering useful information and furnishing it to the ACC, as they were convinced that the communists would provide such information only to the soviet contingent.[30] The
head of the bureau, Sándor Szalai, even aimed at preempting the communists in liaison with the soviet section of the ACC. Months after he began
his activities he wrote with evident satisfaction: "Russians realize that they
have to 'politicize' in our direction. They greatly value the political direction of *Népszava* [the SDP press organ], while they complain a great deal
about the *Szabad Nép* [the communist daily]."[31]

This was hardly a time for interparty squabbling though; the primary
task of every political group was to repair Hungary's image before foreign
opinion. Realistic politicians, Gyöngyösi in particular, knew that the truly
important trial still lay ahead: the making of peace at an international conference where Hungary could count on few, if any, friends. The conference
would in any case merely put the seal of approval on decisions made by a

smaller body, the Council of Foreign Ministers, and within that council the realities of the spheres-of-influence arrangements would prevail.

Hungary was within the soviet sphere and Stalin had said that the borders would remain unchanged. The government knew this but the public did not and large segments of it continued to hope that at this second peace conference within a quarter century Hungary would receive "justice." How could it be otherwise? Three million Hungarians lived outside the nation's borders—What kind of justice could sanction that? Gyöngyösi knew that in the Council of Foreign Ministers this would be a mere piece of statistic. So would be the fact that of all European states Hungary was the only one whose population was smaller than the total number of its nationals on the continent.*

The counts of indictment against the nation were too numerous and too grave to be treated with leniency. Its wartime behavior had been bad enough, but its distant past was not much more creditable. Its social elitism, the shameful treatment of the peasantry by the gentry and the aristocracy, the blatant rank-consciousness of the middle and upper classes, and the sometimes covert, at other times open, anti-Semitism were all blots on Hungary's claim to being a civilized nation.

On the other hand there was the fact that, except in the Swabian German segment of the population, there had never been much enthusiasm for the German alliance. This impassive virtue was however counteracted by the fact that there had been no organized resistance against the occupying Germans, no partisan movement, not even individual acts of sabotage, and many Hungarians had enriched themselves from abandoned Jewish property. Why then should Hungary's ethnic claims be treated with greater sympathy than they had been in 1920? Yet that was what the nation ardently expected.

In the early spring there still was the hope that Hungary might gain merits by joining the crusade against Germany in its final phases. When Pushkin visited Gyöngyösi on April 4, the foreign minister put in a plea for weaponry and supplies so that the new divisions could at last be engaged. He said that he raised this purely military issue because the pleas of the defense minister had gone unheeded. The matter was urgent because if the divisions were not made battle-ready Hungary would be left out of the war.

---

*The census of 1930 showed a largely contiguous centrality of 10.8 million Hungarians; the population of the country itself was but 8.1 million. By contrast 13.8 million Rumanians ruled a state of 18.1 million inhabitants; Yugoslavia, with a total of 11.9 million Serbs, Croats, and Slovenes, had a population of 13.9 million; Czechoslovakia's 10.2 million Czechs and Slovaks dominated a state of 17.7 million people. —New Hungarian Central Archives, Foreign Ministry Files, Foreign Administrative Mixed, Box 171, 9/e.

Pushkin admitted that this was indeed a possibility but added that the new Hungarian army might still have tasks to fulfill. At the end of the conversation he casually remarked that according to information at his disposal the entire territory of Hungary had on that day been liberated.[32]

This in effect dashed the hope of ending the war on the victorious side. (Although Hungary had declared war on Germany on December 31, 1944, none of its units had been involved in the hostilities.) The merits the nation would now place before the great powers would have to be all political. The foreign ministry hastily prepared a document to point out the difference between the composition of the parliament in the Horthy era and of the National Assembly recently formed. Of the 245 deputies elected in 1931 (to the lower house) 84 had been landowners, 41 civil servants (beholden to the regime), 15 bankers, one a manufacturer, 37 lawyers, 12 clergymen and 14 high military officers. As to social rank, 19 had been counts and nine barons.

In the selection of the upper house the electorate had no voice; its membership was in part hereditary and in part derived from ecclesiastical and state functions; the others were co-opted by the members or named by the regent. Horthy had made his own son a life member and for the rest he chose the highest personages from finance, industry, and the landed aristocracy. Members of the cabinets were almost without exception of the same ilk. The agricultural portfolio had always been held by a landed magnate, and that of finance by a prominent banker.[33]

The composition of the new unicameral National Assembly was telling evidence of how Hungary had democratized its system. (As we have pointed out before, the membership had an almost sansculotte character.) What the memorandum did not say was that despite these changes a different kind of elitism had replaced the older one.

The aggressive conduct of the MKP and the dominant position of the soviet delegation on the ACC was causing alarm at home and abroad. There was a larger fear as well. The extensive annexations carried out by the Soviet Union, with hardly an attempt to ascertain the will of the populace, often under the guise of liberation, made it a real possibility that Stalin was determined to enlarge his sphere, preferably while the fluid conditions created by the war lasted.

Another secret and undocumented memo prepared in the foreign ministry noted that the fear of Hungary being absorbed in the Soviet Union ran deep in the country. The precedents of the Baltic States, of the Rumanian provinces of Bessarabia and Northern Bukovina, and of the large eastern slice of Poland were ominous. The memo nevertheless argued that this fear was probably without foundation. The soviets in their grave economic condition had no interest in incorporating any more border states; as a matter of fact if any state made such a request, it would probably be rejected.[34]

Still, current restraint by soviet leaders was no guarantee for the future. Expansion was a permanent feature of Russian history and there had not been a century at the end of which the empire was not substantially larger than it had been at its beginning. The present world was of course a different one; principles of national self-determination could not be ignored altogether.

In the end it would be a matter not of playing by the rules of international conduct but of finding a formula, as the soviets were so adept at doing, by which outright conquest could be made to appear as a response to the national will. Stalin was already rumored to have remarked to Czechoslovak statesmen who had wondered how 600,000 ethnic Hungarians could be expelled from the republic that in the end the problem of Hungary was a problem of enough cattle cars. Would the soviet dictator not apply this practice to troublesome Hungarians when the time came to complete the "liberation" of the country?

Chapter 4

The Plight of the Minorities

Hungarians dated their troubles with their border states from the Trianon treaty of 1920; those states in turn looked back on centuries of oppression by a haughty and arrogant Hungarian ruling class. Such disputes have a rhetorical content, but they are ultimately resolved by the great powers and often with scant regard for the merits of the respective arguments.

Czechoslovakia and Yugoslavia emerged from World War II as victor states; Rumania emerged as a defeated one but with the advantage of having acknowledged its defeat months before Hungary did and having taken an active part in the war against Germany. In the face of these facts the most Hungarians could hope for was that a modest portion of their claims against Rumania would be honored; no claims were made against Yugoslavia, and Czechoslovakia's victor status precluded any land cession by it to defeated Hungary. In its case the dispute shifted from territories to the treatment of the Magyar minority.

In view of Czechoslovakia's sterling democratic credentials it was paradoxical that Hungarians in Slovakia were in a worse position than they were in either Rumania or Yugoslavia. Between the wars minorities in that republic had enjoyed all the freedoms and privileges a truly democratic state would grant to its citizens. The two presidents of prewar Czechoslovakia, Tomas Masaryk and Eduard Beneš, both men of enlightenment, dreamed of turning their republic into a multicultural haven, somewhat on the Swiss model. But Hitler's campaign on behalf of the German minority and the passions it aroused put a drastic end to that dream. As the Czech leaders saw it, their minorities, instead of appreciating the freedoms they enjoyed, turned against their adoptive homeland and made common cause with predatory neighbors. These leaders returned from the war hardened and disillusioned, determined to have nothing further to do with minorities; they would solve the problem by summary expulsions.

In the case of the Sudeten Germans they encountered no major obstacles to this radical measure; anti-German sentiment was so intense in all of Europe that any hardship imposed on those people met with understanding and approval. But the Hungarian minority in Slovakia posed a more thorny problem, mainly because the land they inhabited had been part of "historic Hungary" and its transfer to Czechoslovakia had been controversial from the start.

There are regions of the world where questions of ethnicity are not of serious concern and carry little political weight. East Europe is not such a region. Since ancient times it has been plagued by nationality problems, the deeper roots of which elude the researcher; undeniably though, attitudes of racial discrimination are exceptionally strong. Hungarians had been guilty of such attitudes since the founding of their homeland. The prejudices of the ruling Magyars extended not only to ethnic minorities such as Slovaks, Rumanians, Ruthenians, and Croats, but even to the underclass of their own society, the landless cottars whom they regarded as some lesser breed unworthy of full inclusion in the commonweal. Because this system of dominance and subjection survived for some thousand years, it acquired the status of a natural right. True, the Magyars themselves had been conquered by Mongolians, Turks, and Austrians, but they had never allowed themselves to be subdued and in the end asserted their independence.

The Trianon treaty would have traumatized the nation in any case because, short of the total extinction of a nation's independence (as happened with Poland and the Baltic states), it was unprecedented that a state of a millennial history as Hungary should lose two-thirds of its territory; the self-image of Hungarians as a master race made it even worse. If the Slovaks and Ruthenes in the north had never made an attempt to achieve their independence, by what right did they receive it as a gift? The concept that underlaid the creation of Czechoslovakia was Slavic solidarity—but how could the presence of three million Germans and one million Hungarians in the republic be reconciled with that concept?

For nearly two decades Hungarians hoped that the great powers would acknowledge their mistake and draw the necessary conclusions. Only when all such hopes proved futile did they make common cause with Hitler's revisionist agenda. Such was, in bare outline, Hungary's role in the partition of Czechoslovakia—now it was asked to shoulder the guilt of Munich.

The Czechoslovak government-in-exile in London did not at first give any indication that it held the Magyar minority responsible for the disaster that had befallen the republic. Vladimir Clementis, a member of that government and the man who after the war became the most unbending foe of Hungarians, during the war spoke in terms of warmest appreciation of them. "Anyone who observed the situation before Munich in the Hungarian regions of Slovakia," he wrote in 1943, "can bear witness to the fundamental difference that existed between them and the Sudeten regions. The majority of the Hungarian workers and peasants not only appreciated the political and cultural progress which the republic assured them but had a clear idea of the position of Czechoslovakia in the European situation as well as of its importance for the hopes of their brothers in Hungary and thus for the whole Hungarian nation."[1] At another time he wrote: "The energetic actions of Hungarian democrats inspire in us the highest hopes for the future, even if a portion of the 'intelligentsia' succumbs to the influence of retrograde Hungarians."[2]

It was this retrograde Hungarianism, condemned not only by the Slovaks but by Hungarian progressives as well, that stood in the ideological center of the debate. "Retrograde" was a synonym for "reactionary" and of all the reactionary sins irredentism was the worst. The communists regarded it as an expression of a vaguely conceived national greatness that took precedence over social improvement; its proponents were ready to condone grossly inequitable conditions inside the country as long as the country was large enough to satisfy their racial megalomania. Jozsef Révai, the chief publicist of the party, chided his compatriots: "It is one of the diseases of Hungarian public life that we judge world politics provincially, on the basis of narrow local viewpoints. . . . We are inclined to believe that the world turns around us; consequently we judge great power realities of international politics according to our [national] wishes."[3]

The message, particularly galling to Hungarians, was that they should subordinate their national aspirations to larger international interests, which, by communist logic, meant the interests of the Soviet Union. This at a time when soviet foreign policy was dictated by a ruthless egoism that could only partly be justified by the claim that the experience of the two world wars showed that Russia needed "friendly" states on its borders. The Hungarian Communist Party (MKP) was never able to explain how a prosoviet orientation of Hungarian foreign policy would redound to the benefit of the nation. Even decades later communist historians could offer only generalities. "The communist party based Hungarian foreign policy on the premise that cooperation among the great powers was possible and . . . it judged Hungarian-soviet relations to be of particular importance, not because of selfish party interests, but because of the realization that our national interests, political, economic and foreign relations demanded the cultivation of Hungarian-soviet cooperation."[4] Why this should have been the case was never explained.

What proved extremely embarrassing to domestic communists was that in all of Hungary's disputes with neighbor states the soviets supported the opposing side. The only way to gloss over this fact was to relegate nationality questions to secondary importance, which might have been exemplary Marxism, but had a negative propaganda value. The Smallholders were almost shamelessly western-oriented. The United States was the center of their hopes, because of its wealth and generosity and because it was not weighed down by the legacy of Munich. It had a greater sympathy for the fate of national minorities than did any of the European powers. (As a matter of fact when the Committee on Postwar Programs in the State Department made its recommendation regarding the frontiers of Czechoslovakia, it suggested that the United States "should favor cession to Hungary [of a region in South Slovakia], either on the basis of direct negotiation between Czechoslovakia and Hungary, or on the basis of a determination by appropriate international procedures."[5]) On the question of the Rumanian border too the western position was more favorable to Hungary than the one laid down by Stalin before armistice negotiations even started.*

The wartime pronouncements of men like Clementis gave Hungarians reason to believe that the Magyar minority in the republic would be treated with the greatest forbearance. This hope was disabused from the start. As soon as Slovak authorities returned in the train of the advancing soviet armies to the lands that had been joined to Hungary in 1938, the persecution and expulsion of Magyars, often accompanied by the confiscation of their property, began at once. Hungarians might have thought of this as a series of extralegal measures that would be discontinued once a central government had been established in Prague. But Prague at this time was still under German occupation and the government-in-exile in London was powerless. Decisions concerning minorities were being made in Moscow.

---

*It should be noted that Stalin had an odd compensatory sense in territorial questions; he was partial to countries he had taken land from in their disputes with other states. Poland, Czechoslovakia, Rumania had all ceded territories to the USSR, but Stalin made sure that they were protected against the claims of other countries.

President Beneš, with a delegation of Czech and Slovak politicians, arrived in the soviet capital on March 17, 1945 and immediately began negotiations with Stalin and Molotov about the principal lines of Czechoslovak internal policy. Beneš later reported to U.S. Ambassador Averell Harriman that his government had been authorized to exercise control within the republic's pre-Munich frontiers and added that: "The question of Ruthenia would also be settled after the war depending on the will of the people." The meaning of this last statement, the result of negotiations with Stalin, was obvious. Harriman reported to the State Department that Beneš "did not seem particularly exercised over the possibility of losing Ruthenia."[6]

While in Moscow, Beneš was joined by a group of Slovak communists and democrats who constituted themselves as the Slovak National Council. Benes told Harriman that he intended to grant the council home rule in Slovakia.[7] It was headed by one Gustav Husak, a communist with uncharacteristically strong national convictions who was dedicated to the proposition that the republic should become a purely Slavic state. It was most probably at the behest of Husak that Beneš sought Stalin's sanction for the expulsion of Germans and Hungarians from Czechoslovakia. According to Harriman's report, the soviet dictator authorized the expulsion of "about 2 million of the 3 million Germans within Czechoslovakia . . . and similarly 400,000 of the 600,000 Hungarians."[8]

Presumably the Hungarian communists knew nothing of Stalin's collusion with the Czechoslovak government and continued to affirm that close cooperation with the Soviet Union was in Hungary's best interest. In fairness to them, even if they had been privy to that collusion they could not possibly have advocated a different course without dismantling the concepts that held their policy together.

On April 4, 1945 (the very day when all of Hungary was "liberated"), the head of the Czech Communist Party*, Klement Gottwald, in a speech in the Slovakian town of Košice—which for the duration of the war served as the capital of the republic—outlined among other things the nationality policy of the government. The undiplomatic harshness of his speech made it clear that he was speaking with the sanction of an authority that could not be gainsaid. "In our relations with Hungary," he said, "we will utilize to the fullest the situation [created by] the armistice which, thanks to the help of the Soviet Union, had secured Czechoslovakia such [significant] advantages." Without elaborating what measures he had in mind, he added, "Later,

---

*There were two communist parties in the republic: the Czech, which was one of four political parties in Bohemia-Moravia, and the Slovak, which was one of two parties in Slovakia after it had merged with the Social Democratic Party.

after the injustices and wicked acts of the Hungarian occupiers had been atoned for, the government will support endeavors aiming to bring a new and truly democratic Hungary . . . closer to us neighboring Slavic nations and states."[9]

That however lay in the incalculable future. For now the so-called Košice Program governed the position of Magyars in the republic and it was unequivocal:

> (1) "Only those residents of Hungarian nationality will retain Czecho- slovak citizenship who were anti-Fascist or who participated in the resistance movement for the liberation of Czechoslovakia or who were persecuted for their loyalty to the republic; (2) the Czechoslovak citi- zenship of all other Hungarian residents is withdrawn, but they will be given the opportunity to opt [between staying as non-residents or moving to Hungary] and every appeal of this nature will be separately examined; (3) those persons of Hungarian nationality who have com- mitted a crime against the republic or against other nations, especially the Soviet Union, will be placed before a tribunal, deprived of their citizenship and forever expelled from the territory of the republic."[10]

The program attempted to put a veneer of legality over another and more radical one prepared by the Slovak Communist Party in February of that year, which called for "a new conquest of the fatherland." The manifesto that announced the program read in part: "The Slovak peasant and worker who had been squeezed out of the rich southern regions and for centuries had lived in subjugation in the mountains, must be returned to the old Slo- vak areas and given the opportunity for a decent human life."[11]

The position of the Magyar minority in another lost province, Transylvania, differed in specifics from the one that obtained in Slovakia but was similar in general terms and had different roots. Here too there had been changes in sovereignty within a short period of time: all of the province had passed to Rumania in 1919; the northern part of it was, thanks to Hitler's largesse, returned to Hungary in the summer of 1940, only to be lost again in the closing phases of World War II. These territorial changes activated intense emotions of fear, anger, heady triumph, and a desire for vengeance.

The fate of Transylvania, more than that of any other province, illus- trated a peculiar weakness of the peace settlement after the First World War. When the peacemakers attempted to draw borders along ethnic lines, making that in fact the principal criterion, they went against centuries of European tradition. No major peace treaty on the continent in modern times had accorded ethnic concerns any weight. The Peace of Westphalia, the Peace of Utrecht, the provisions of the Congress of Vienna, and even those of the Berlin Congress of 1878 at a time when nationalism was a potent force, all ignored the question of what language was spoken in what province. Power realities alone mattered. When during World War I President Woodrow Wilson introduced the concept that each nation, large, small, or minuscule, should rule the land it inhabited, he did not suspect that the application of the principle, far from settling long-standing conflicts, would exacerbate them.

The fact that the ethnic lines were often whimsically drawn proved only part of the problem. The more important part was that efficacious governance required more than ethnic homogeneity. Tradition was a great unifying force, so was economic self-sufficiency, and the latter could often be achieved only by creating an ethnic mix; finally there was the question of the creative and organizational talents of one nationality compared with those of others. The Dual Monarchy, jointly and severally, had constituted a well-balanced economic unit and no one could deny that the respective territories were efficiently governed (even in the troublesome South Slavic regions); the Paris peace settlements destroyed that unity.

Perhaps there was some merit in the argument of a school of economists who held that the Great Depression, in Europe anyway, had its roots in the breakup of the Dual Monarchy and the intense economic nationalism that followed. But this argument, even if proven, could not override the liberal conviction that every nation, whatever its historic credentials, deserved a government of its own free choice.

Between the two wars the League of Nations Convenant provisions governed the treatment of national minorities, but the experience was an unhappy one. The fact was that a supranational organization with no effective means of enforcement could not breach the sovereignty of a state and thus each governing nation decided for itself how it should treat its ethnic minorities. Naturally, even when the treatment was good the discontent of one national group governed by another, especially when the governing group had been the subject nationality not long before, proved an unsettling factor. The Germans provided the most flagrant example of this situation; after World War II nations with a German minority simply expelled these troublesome foreigners, with the result that some ten million Germans were driven, or went on their own, to the western occupation zones of Germany.

To a lesser degree, Hungarians in Czechoslovakia, Rumania, and Yugoslavia suffered a similar fate. For months in the closing phase of the war and even after the return of peace Hungarians at home watched helplessly as their ethnic kin suffered indignities of the worst kind in neighboring states.

In Slovakia the persecutions were carried out within a legal framework. But Northern Transylvania was infested with lawless bands, the most notorious of which was the so-called Maniu Guard, rabidly nationalistic Rumanians, who waged a war without rules against the Hungarian minority. One source summarized events in this brief account: "After a sendoff at a festival which combined [nationalistic passions] with a religious service . . . in Bucharest, the Guardists marched into [Transylvania] and under the pretext of restoring order inflicted a bloodletting on Northern Transylvanian Hungarians that even a nightmare fantasy could not exceed. Hungarian peasants were beheaded with axes, women and infants were murdered."[12]

When the Allied Control Commission (ACC) for Rumania, sitting in Bucharest, learned of these atrocities, it ordered that Rumanian administration be withdrawn from the province. In effect the allied powers did not recognize Rumanian sovereignty over Northern Transylvania and even took the position that Rumanian participation in the war against Germany was an unofficial affair. Only in October 1944 was there formed in Rumania, on the Hungarian model, a National Democratic Front, uniting sober and responsible political elements; its draft program called for an equality of rights for minorities. It also demanded a strict observance of armistice terms; not until the spring of 1945, however, was a measure of legality and order returned to Northern Transylvania. On March 6 a moderate politician, Petru Groza, formed a government; a month later Stalin, in his capacity as soviet premier, authorized the reintroduction of Rumanian administration into Northern Transylvania.

Thus, just when in Czechoslovakia the persecution of minorities began in earnest, in Rumania the Groza government made sincere, if not always successful, attempts to normalize relations between Rumania and Hungary. On May 16 Groza journeyed to the Transylvanian capital, Cluj, and made a speech that, although more declamatory than businesslike, made a favorable impression on Hungarians and Rumanians alike. He promised that all anti-Hungarian agitation would cease and that Transylvania would be free of chauvinism and "will give bread to both Rumanians and Hungarians."[13] In an interview with the Hungarian National Union (an organization sponsored by the Rumanian government and perceived by most Hungarians as collaborationist), Groza said: "Transylvania to [Rumanians] cannot be a wall, only a bridge to Hungary. . . . It is in our common interest that [we] go to the peace conference in mutual agreement."[14]

He was asking for the impossible. No province stood closer to Hungarian hearts than Transylvania; to renounce it would mean betraying a national trust. On the other hand no Rumanian statesman could hope to be left in office if he relinquished any part of the province to Hungary. The question ultimately would have to be resolved by the great powers, more precisely by the one power whose military forces occupied the province. In Yugoslavia too ethnic Hungarians suffered grievously at the hands of largely Croatian forces. There were murders and expulsions; "četnik" units even pushed into Hungary proper, occupied several localities, and declared them joined to Yugoslavia. Only when Josip Broz Tito's government finally established itself in the fall of 1944 did the outrages gradually cease. There was no open territorial question between Tito's government and the Provisional Government in Budapest—the Hungarians did not lay claim to any territories, Belgrade reciprocated in kind, and in this direction, too, relations were normalized.

Only Czechoslovakia was unrelenting. The Yugoslavs made no attempt to exploit their victor status beyond preserving the territorial status quo; the Czechoslovaks translated it into license to flout international law and comity. Protests were of no avail; Stalin had given his assent to the persecutions and expulsions. As a matter of fact, during a toast offered at a dinner party in honor of President Beneš he was reported to have said that the Communist Party of the Soviet Union (CPSU) had "reoriented" its position on nationalities to meet present conditions. As Harriman quoted the speech to the State Department: "The various communist parties would become nationalist parties interested in the national interests of their own countries."[15]

The political committee of the MKP (equivalent of the soviet politburo), under Rákosi's chairmanship, discussed the vexing minority question at one of its sessions in July. This committee, composed of the leading figures of the party, had in effect become a parallel government and by far the more influential one of the two. The topic of the discussion was the outcome of trips two members of the committee, Rákosi and László Rajk, the former as deputy premier, the latter as minister of the interior, had made, first to Belgrade then to Prague. The intention had been to discuss the problem with their communist comrades who also supposedly viewed it through the broad and crystal-clear lens of Marxism.

The first trip confirmed what had already been settled, that the Hungarian minority would be reintegrated into the Yugoslav citizenry. The Prague trip on the other hand proved a fiasco. The Czechoslovak leaders, communist or otherwise, confident that all great powers would support their position, turned a deaf ear to Hungarian entreaties. They refused to reconsider their decision to expel their ethnic Hungarians, even those who had been born in the republic; three days after the departure of the delegation, over the signature of President Beneš, they issued a decree that, in pursuance of the Košice Program, deprived all Hungarian nationals in Czechoslovakia of their citizenship.

Here was another arbitrary act that called for Stalin's intervention; pleas to any other authority would have ended up on his desk anyway. Rákosi, who during his exile had grown quite close to Stalin, was enjoined by the political committee to address an appeal to the dictator. His letter is a good example of the casual and superficially intimate tenor of correspondence between comrades.

> To the Generalissimo of the USSR, to Comrade Stalin!

> I know how busy you are, but an important question has arisen here which we do not want to decide without consulting you....

> The Czechoslovak government and Communist Party decided to resettle all Hungarians from the territory of Czechoslovakia. They argue that they have to create a strong Greater Slavic federation and the presence of the 600,000 Hungarians in Czechoslovakia complicates matters. Their main contention is that you are in accord with this plan. When the Czechoslovak delegation visited you at the end of June and broached the subject of whether they can treat Hungarians in the same way they treated Germans, you have allegedly reacted positively and even said, "Let them get what's coming to them."

> Comrades from the Czechoslovakian Communist Party with whom I talked two weeks ago in Prague added to this argumentation that their party has already gone so far in this direction that it cannot turn back without suffering great loss. We asked Czechoslovakia to treat its Hungarian minority as it is treated in Rumania and Yugoslavia. (Comrade Tito, with whom I talked about this, supported [my] request.)[16]

The letter closed with stock phrases and the assertion that if Hungarians were expelled from Czechoslovakia, it would be a terrible blow to the young Hungarian democracy and would strengthen the position of the chauvinists. There is no record of an answer to this letter, either in the archives or in extant published histories.

Already Rákosi was the most authoritative voice both in and outside of government. His correspondence, most of which was preserved in the archives of the Institute of Political History, contains letters directed to him and answered by him on subjects of the most varied nature; they leave no doubt that every major decision in the party or the government was provisional until he had given his approval. At the same time foreign policy items took up only a minuscule portion of the correspondence. In Rákosi's pronouncements, comments on international politics were brief and perfunctory, with a predictable tendency to judge them in purely Marxist terms. At the June 7 meeting of the political committee of the MKP he offered a dogmatic and depthless observation on the developments of great power relations.

In connection with the destruction of German fascism [he said], friction has increased among the great allies. The force of cohesion has ceased. Problems have come into the foreground. Sanguine hopes based on the falling-out between the Anglo-Saxons and the soviets [proved to be] like the hungry hog dreaming of corn. For now, not the soviet-Anglo-Saxon differences are the greatest, but those among the imperialists themselves.[17]

Rákosi's relative lack of interest in foreign affairs was a misfortune because he alone, by dint of his exceptional intellect and his prestige at home and in Moscow, could have introduced into Hungarian diplomacy an element of Realpolitik that was essential in order to reconcile the widely different expectations across the political spectrum. Apparently though he looked on foreign policy as a tiresome if necessary concomitant to the attainment of a global socialist society. His lengthy and unsparingly candid letters to his mentor, Georgy Dimitrov, contain practically no reference to foreign affairs. One recurring exception was his dealings with the Czechoslovak "comrades." But he saw the negotiations with them not as attempts to minimize the increasingly hostile relations between the two countries but as a case study of how problems that bedeviled ordinary diplomats could be solved in a spirit of comradely understanding. "We told the comrades," he wrote at one point to Dimitrov, "that the expulsion of Hungarians from Czechoslovakia would be an enormous political blow to [our] democracy that is not too strong anyway and how it would strengthen not only the Hungarian reaction but all those forces which are opposed to the development of a strong, healthy democratic bloc." Elsewhere he countered the claim of his vis-à-vis that in taking the anti-Hungarian measures the Czechoslovaks were aiming at the creation of a greater Slavic confederation by saying that such a federation could be strong only if it gained the confidence of the Hungarian (and Rumanian) democracy.[18]

All the while Hungary's 8.5 million subjects were starving and Rákosi could not be blamed for being far more preoccupied with questions of the food supply than with the vast imponderables of global politics. "We stand before a harvest," he wrote to Dimitrov in June (the letter is undated). "Our food supplies are exhausted and at the same time we have to produce huge quantities to feed the Red Army. . . . This is one reason for the rapidly growing inflation and the deterioration of the food situation. Our bread ration is still only 100 grams a day and among workers and small clerks there is a veritable famine." He added bitterly that in Vienna and Berlin the rations were much larger, yet much of Hungary's grain had been taken to Austria to feed the soviet forces there.[19]

He never went so far as to be critical of soviet policy, but the victimization of Hungary evidently greatly disturbed him. He must have fought passionate battles with his communist conscience trying to convince himself that the suffering of his people was a necessary prelude to a world revolution in a later generation. That might have been one reason why he turned away from foreign policy questions—dealing with them would inevitably open his eyes to certain truths he preferred not to contemplate. Stalin might well have been right in predicting that the crisis of capitalism was inevitable, but given the realities of international economics, that too was a prospect to be witnessed only by later generations.

Chapter 5

Swabians and Jews

The developing antagonism between the western powers and the Soviet Union was fairly accurately reflected on the various Allied Control Commissions whose task it was to supervise the fulfillment of the armistice terms in the defeated states. These commissions met periodically but in each case their collective authority was circumscribed by the commanding position of the occupying power. In Italy and Japan the Americans effectively excluded the soviet representatives from decision-making; in Rumania and Bulgaria the reverse was the case. In Finland, where there were few occupying troops, the ACC worked smoothly and complaints were rare on either side.

In Hungary the picture was mixed. Klementi Voroshilov was often high-handed but when confronted with firmness he could be amenable, not only toward the western allies but toward the Hungarians as well. More than any other country in the soviet sphere, Hungary seemed to be the testing ground of the extent to which the western powers could prevent the soviets from dictating the policies of a country they occupied. All evidence seems to indicate that Stalin was still undecided as to whether to integrate Hungary into the soviet bloc. While, as in other countries in his sphere, right-wing parties were excluded from the political process, those in the center, as well as the left, were allowed ample scope and the Hungarian Communist Party (MKP) benefited only marginally from the presence of soviet soldiery. Members of the Provisional Government routinely consulted soviet authorities but the latter did not exert undue pressure for the enactment of specific measures.

With the war at an end the role of the ACC diminished perceptibly, but on balance this worked out to the detriment of the western powers. Certain important areas of activity, such as Hungarian airspace and joint enterprises with large foreign investments, remained under effective soviet control. This was the pattern in all countries under Russian occupation and the western members of the ACC complained repeatedly. The American chargé in Moscow, George Kennan, wrote to Ambassador Averell Harriman, who was then in Washington; "our only hope of getting anywhere [in gaining an equal voice on the ACCs] would be to make up our minds that if we do not get full tripartite treatment we will withdraw not only from participation in the Control Comission but our political representatives as well."[1] This was rash advice and Washington ignored it; German surrender was only one week old (Kennan's telegram was dated May 14) and allied cooperation in making peace was as important as it had been in making war.

The German question had long been a preoccupation of international policy makers. After World War I the question had been how to punish Germany; after World War II the question was how to punish Germans. Millions living in foreign countries had proved such fanatical adherents of Hitler's program and responded so eagerly to his calls for pan-German solidarity that they became fifth columnists in their embattled homelands. Now the backlash came in a powerful wave.

Stalin deported most of the Germans dwelling along the lower Volga; the Poles, whose national territory had been shifted some one hundred miles to the west, found themselves in possession of regions that had previously belonged to Germany and forced most of the German inhabitants to flee. Czechoslovakia, as we have noted, was in the process of doing the same. Hungary, when the war ended, had not yet decided how to treat its own German-speaking citizens. Opinions were divided, as they were bound to be when a whole body of people was to be judged collectively.

The German minority, called Swabians though they were of mixed origins, had lived in the country for centuries (some since as far back as the Thirty Years' War) and had been on the whole a peaceful ethnic group, in general not prosperous enough to give rise to jealousy or resentment. When Hitler began his campaign on behalf of the *Volksdeutsche*, Germans living outside the Reich, Hungary's Swabians bestirred themselves, as if they had suddenly discovered their superior racial heritage. Many who years before had changed their names to Hungarian ones reverted back to their family names. A *Volksbund* of German-speaking people was formed, stressing ethnic loyalty to Germandom over their Hungarian citizenship. But this was an unimpressive lot; many were economically poor and culturally deprived, certainly not the type to attract more respectable members of the German ethnic community. At the same time hundreds of thousands of Swabians remained steadfast in their loyalty to Hungary, and many of these joined the Fatherland Front, which stood largely under Catholic influence.

One week after the country's liberation the cabinet met to discuss the Swabian question. Some ministers took the simplistic position that all *Volksbund* members were fascists and had to be dealt with accordingly; others were inclined to be more discriminating. The most vocal champion of a harsh policy was the spokesman for the National Peasant Party (PP), and for a plain enough reason. The PP was a party of small landowners, many of whom were bitterly disappointed with the results of the land reform, which left some of them landless and others with impossibly small allotments. The daily organ of the party claimed it was speaking "from the heart of the nation" when it called for the summary expulsion of the "Swabian traitors."[2]

But the matter proved complex. A survey of the wartime conduct of the Swabians showed that those most guilty of political misbehavior were usually also the least prosperous, thus little was to be gained in terms of land by their deportation. The PP plan thus called for the deportation of every person who, at the last census, identified himself as a German—only those who had rendered signal service to the country were to be exempted. But the review of the census results led to further confusion. In 1941 altogether 300,149 persons had declared themselves to be German nationals, while another 477,057 listed German as their mother tongue. In 1940 roughly 40 percent of all German-speaking persons were members of the *Volksbund*. But these figures pertained to Greater Hungary after the Vienna Awards and there were no reliable data for Trianon Hungary. Also, some 60,000 to 70,000 Swabians had left with the retreating *Wehrmacht*, some later returned but others remained abroad, mainly in Austria. Thus, while a comprehensive deportation was rejected by most parties, none of them really had any idea how to carry out a selective one.

The cabinet was in any case too small a body to settle the question, and on May 14 representatives of the coalition parties met in conference to discuss it. The group adopted the virtuous principle, "In Hungary, there is no Swabian question, only a question of German fascists," but then went on to adopt a resolution calling for the deportation of former members of the Schutzstaffl (SS) and the confiscation of the lands of *Volksbundists*. In general though the consensus was that a comprehensive Hungarian solution of the Swabian question would have to await a general European settlement of the question.[3]

The routine use of the word "fascist" in such connections suggests that the soviets were the moving force behind the expulsions. We have seen that Stalin raised no objection to the removal of offending ethnic minorities from Czechoslovakia. The western powers were ambivalent in their position, but the United States rejected the principle of collective responsibility. The Czechs and Slovaks meanwhile were taking advantage of the unsettled conditions created by the war to deport as many undesirables as their police forces and means of transportation could handle.

One reason why in Hungary the interparty conference in May took a stiffer stance on the Swabian question than the cabinet had was that the continued expulsion of Hungarians from Czechoslovakia made the problem of settling them increasingly urgent, lest they became public charges. And so, shortly after the conclusion of the May conference, the Provisional Government sent a note to Moscow, informing the soviet government that "the calling to account of Germans in Hungary who are permeated with the fascist spirit is in progress. We have come to the conclusion that the Germans who had been in the service of Hitlerism should be removed [from Hungary]. Only in this way can we ensure that German spirit and German oppression will not come to dominate the country." The note respectfully asked that 200,000 to 250,000 "such fascists" should be allowed to be deported to the soviet occupation zone of Germany.[4]

The western powers did not receive a similar request; they were less likely to be receptive than the soviets. As a matter of fact the Americans on June 10 (a probable date: Washington's note to Prague was forwarded to Budapest on June 12) emphatically raised their voice against such collective action. In an uncharacteristically stiff note Washington dissociated itself from the mass deporation of Germans that the Prague government was contemplating. It insisted that when such action was taken, the interests of not only of Czechoslovakia but also of European peace and security should be considered. The process was furthermore expected to conform to the principles of international justice and to be in any case gradual. The note made it clear that American reservations applied to deportations from Rumania, Yugoslavia, and Hungary as well but referred to the case of Czechoslovakia in particular, stating that while there undoubtedly were criminal elements among the ethnic Germans, the United States government "would not consider it justified to endeavor to deal with all members of an ethnic group who constitute a minority, as criminals . . . [simply] because of their ethnic origins."[5]

In Hungary meanwhile the ponderous process of distinguishing between "good" and "bad" Swabians and setting criteria for treating each had got under way. A special government decree of July 1 called for an examination of the past conduct of Germans living in Hungary and provided penalties for those who had been "active Hitlerites," those who had been members of Hitlerite organizations, and finally those who, although not members, had associated themselves with the aims of such organizations. (Germans who had neither been members of the *Volksbund* nor supported it were unaffected by the decree.)

In theory the measure ensured a system of orderliness but in practice it led to confusion and predictable excesses. What made the process particularly touchy was that these ethnic Germans had no native government to speak for them, no advocate before international forums. There was moreover no expert body to determine how many expellees Germany was capable of taking in. At least Hungary did not follow the example of Poland and Czechoslovakia in undertaking indiscriminate deportations but allowed voices to be raised against the inhumanity they often entailed.

The most weighty voice was that of Cardinal József Mindszenty, the man who in the years to come would wage a dignified, if often ill-tempered, campaign against the communist-dominated regime. He was a Catholic of the old school, a confirmed royalist, a man of great personal courage but bent on becoming a martyr, with a deep social conscience, though by no means a champion of the underclass. He was a fervent opponent of land reform and a sworn foe of socialism in all its varieties. He early became a thorn in the side of the Provisional Government, a hopeless and destructive reactionary in the eyes of the communists. Himself a Swabian (his family name was Pehm), he was determined to save all but the worst of ethnic Germans from punishment. (Always active in preventing persecution, he had been instrumental in saving a number of Jews during the German occupation.) He probably would have raised his voice in defense of the Swabians in any case but he was particularly exercised by every measure that emanated from the current government, which, as he saw it, was communistic in all but name.

No sooner did the action against the Swabians begin than he wrote to Premier Béla Dálnoky-Miklós in protest. Loyal Swabians, he charged, were being lumped together with *Volksbundists* and treated like criminals. When he had no answer to his letter he wrote a second one, calling the premier's attention to the internment of men and women, many old or sick, and to the children who were left behind when their parents were deported—children whom no one dared to protect for fear of political consequences. The three-member commissions, he wrote, who were empowered to decide which of the Swabians were subject to property loss and deportation, appeared in communities and passed down judgments out of hand. Their determinations were based on such factors as where the parents of the persons in question were born, what language they spoke at home, and whether they spoke Hungarian in a faulty manner. Those found culpable were put up in cramped quarters. The people who came in their place showed little interest in working the land; they slaughtered the livestock, ate the meat, then returned to their old homes.

All this, Mindszenty insisted, was against Christian principles as it was also against the principles of a constitutional state. (He quoted Canon No. 2219, paragraph 3, which states that a person cannot be held responsible for the crime of another.) What had happened, he conceded, was that a part of the German population in Hungary had for a decade and a half lost its reason. "We condemn that," he wrote. "But it is also true that the Germans of this country have for two centuries been the most adaptable minority, most suited for Magyarization. Now we turn them into enemies, just as we have done with a good part of the Jews." The anti-Swabian measures, he concluded, had been hastily conceived and carried out. He begged the premier to review them.[6]

His was a voice in the wilderness. The process of selection and deportation continued; on the other hand the commissions worked so slowly that injustice was meted out in small measures, and the number of those guilty was a good deal smaller than had been anticipated. In fact the longer the process continued the more its effectiveness diminished. The soviets made repeated representations for greater speed but were powerless in the face of Hungarian reluctance and inefficiency.

The public at large knew little of the persecutions; the affected areas were fairly compact and the press seldom reported the deportations. Every time the matter did come up, inevitably comparisons were made with the fate that had befallen Hungary's Jews—Mindszenty remarked in one of his letters that the previous year Hungary had a Jewish ghetto, now it had a German one. But the Jewish question was bound to come up sooner or later anyway.

The nation was beginning to awaken to the horrible truth that the entire Jewish population in the countryside had been deported in the most inhuman conditions, not to perform labor in "eastern territories" as Nazi propaganda had claimed, but with the ultimate goal of total extermination. Less than 15 percent of the deportees had returned; those who did, often came back starved, diseased, and enfeebled. The armistice obligated the government to release all persons incarcerated because of their racial characteristics or religious beliefs; this provision was on the whole faithfully carried out. Attempts to ascertain how many Jews had perished in the conflagration remained tentative and were revised several times. The number of Jews in Hungary after the addition of southern Slovakia, Northern Transylvania, Ruthenia, and the Bačka was put at 914,000; of those 304,000 lived in Budapest. The number included those who, although formally Christian, were classified as Jews on the basis of Hitler's racial laws. Of the total number 42,000 had been drafted into the labor service. Twenty thousand civilians had been deported from Ruthenia as early as 1941. After the German occupation in 1944, 546,000 were deported from the provinces. Subsequently, under the Szálasi regime, another 181,000 were deported from Budapest. Including the draftees, that made a total of 789,000 deportees. An undetermined number had fled abroad or had been in hiding.[7] By the end of 1946, when most surviving Jews had returned, there were in Trianon Hungary a total of 243,000 Jews—more than twice the number that had been counted the year before. (In subsequent statistics the numbers in every column continued to vary and they were further complicated by the departure, after the war, of tens of thousands of Jews who were largely unaccounted for.)

The Holocaust, which remains the most appalling chapter in the history of World War II, was carried out in a particularly summary fashion in Hungary. Within three months after the Germans occupied the country in March of 1944 they succeeded in doing what in other countries it took years to accomplish: the deporation of over half a million men, women, and children from the provinces; only Regent Horthy's intervention prevented the deportation of the Budapest Jews as well. Some argued later that this proved the at least passive complicity of Hungarians in Hitler's murderous program; others that it was precisely the abnormally accelerated pace of the deportation that prevented the formation of resistance against it. Hitler had by this time realized that the great war he was fighting was lost; he wanted to bring at least his other war, the war against the Jews, to a victorious end. Thus the question of responsibility remains an agonizingly open one.

In the immediate aftermath of the war the foreign ministry did what it could to soften international condemnation. Its most cogent argument was to compare Hungary's attitude toward its Jewish citizens with that of neighbor states. This was judged to be particularly important as certain individual Jews with international connections were broadcasting in print and on the air allegations that Hungary had treated its Jews worse than had other states in the area. The ministry then circulated a study (presumably to be used by legations abroad) that amounted to a vigorous denial. It pointed out that

Slovakia (which now presented itself as a victor state morally superior to Hungary) had introduced sweeping anti-Semitic measures as early as 1939, shortly after becoming "independent" under German protection; that it drew up a "codex" that served as a model for anti-Jewish legislation in other countries; that it confiscated Jewish property and deported its Jewish citizenry in stages. The study further documented that the same procedure was followed in Croatia, and later in Serbia (both under German occupation), and was so deviously carried out that the ultimate destination of the deportations was never established; most Jews simply disappeared. A German officer, when asked what happened to the Jews in the Balkan countries, replied, "Ich glaube die meisten sind schon liquidiert." (I believe most of them have been liquidated.)

As the study showed, Rumanian propagandists were particularly active in trying to whitewash their country's reputation. The actual truth was that Rumanian anti-Semitism had a long history of which the country's own holocaust was a logical consequence. Several times during the nineteenth century Rumania had to be admonished by foreign powers to accord its Jewish citizens full civil rights. But such pressure only increased antagonism toward the Jews. The 1930 census had listed 728,000 citizens of the Jewish faith; by the end of 1943 only 370,000 were left. The internment camps of the pro-Nazi dictator Ion Antonescu took a frightful toll. When, in the first phase of the war against Russia, Bessarabia and Northern Bukovina were occupied by Rumanian troops, 100,000 Jews were killed and buried in mass graves.

In Hungary by contrast there was none of that. The country was for many years a haven for Jews fleeing from elsewhere, until the very moment of German occupation. And when the Provisional Government was established at the end of December 1944, one of its first acts was to declare every public servant who had initiated legislation "that greatly infringed on the national interest, or who has deliberately participated in its enactment" guilty of "anti-national crimes." For such participators the highest penalty was ten years in a penitentiary; if the action resulted in the death of a Hungarian citizen, the highest penalty was death. (Although the act did not specifically refer to the victimization of Jews, that was its intent and most people convicted under it were found guilty of that crime.)

In Hungary, the memorandum finally noted, the persons found guilty, quickly got their punishment, in many instances the rope. In Rumania there were only a few halfhearted trials, some of which resulted in death sentences, none of which however were carried out. The king, at the government's recommendation, granted clemency even to the worst offenders.[8]

A document of this thoroughness and length was obviously prepared not in the first place to counteract exaggerated tales bandied about by certain individuals but in the anticipation that the conduct of each nation vis-à-vis its Jews would figure prominently in the drafting of the peace treaties. To the relief of some and disappointment of others, this turned out not to be the case. Nowhere either in the preliminary deliberations in the Council of Foreign Ministers or at the conference itself was the Jewish question dealt with as a separate substantive item. Perhaps there was apprehension, most likely well-founded, that if the question was placed on the agenda it would bring a flood of recriminations, innuendoes, and denials that would distort the larger political problems the peacemakers had to deal with in building a new Europe. Only in connection with the escalating violence in Palestine and the increasing militancy of Zionists did the Jewish problem become an item of international concern. In Hungary official policy, especially that of the MKP, was deliberate silence; once the legal rehabilitation of Jews and the settlement of their estates was completed, the Holocaust became, for official purposes, a nonevent.

were remarkably hard. "He is completely successful in preventing an indication of emotion or reaction appearing while listening. When talking he can, at will, produce an almost spectacular smile but his eyes do not lose their coldness. His speech is low and quiet yet distinct. He appears to have a determined ruthlessness verging on the fanatical."[1]

The British representative was Alvary Gascoigne, the only one in an otherwise "colorless assembly" of Britishers who, according to a confidential memo by an unnamed communist personage, could be classified as of the first rank. Even that was due "not to his own gifts but to the fact that he was a university classmate of Eden." It was, according to the report, due to his efforts that the British mission did not meddle in Hungary's internal affairs.[2] The American representative, Arthur Schoenfeld, did not arrive until the government had transferred its seat to Budapest in early May. He had been preceded by one Leslie Albion Squires, who had arrived at Debrecen with the title of Secretary and had instructions to seek out proper soviet authorities and make clear to them that he and his colleagues were "members of the staff of U.S. Representative Schoenfeld who will arrive in Hungary in the near future."

Schoenfeld had been chosen for his post as far back as December 2, 1944, but President Roosevelt did not authorize his letter of appointment until January 20, 1945 (the day of the presidential inauguration and the signing of the Hungarian armistice). As a U.S. envoy in Helsinki during the Winter War of 1939–1940 between Finland and the USSR, Schoenfeld had gained the confidence of the soviet government as a coolheaded and moderate man.[3] His status differed from that of Pushkin and Gascoigne in that he was not a member of the ACC contingent but "the principal representative of the United States Government." In his letter of instructions Acting Secretary of State Dean Acheson impressed on Schoenfeld the need to inform everybody concerned that his presence should not be construed "as the opening of official diplomatic relations between the United States and Hungary."[4]

The establishment of such relations would have to await the conclusion of the peace treaty. But the State Department, or rather its Southeast European Office, was discomfited by the fact that the former German satellites had no representation in Washington and their governments and public were consequently uninformed of American reactions to developments in their countries. The office felt that such isolation would lead to an increase of soviet influence. Thus it suggested that the United States propose to send missions to Budapest and Sofia (not to Bucharest, where in the view of the State Department the government had been established by

Chapter 6

Diplomatic Prelude

As a country technically still at war, Hungary could not have diplomatic relations with its former enemies; at the same time nothing prevented the victor powers from posting representatives at the seat of the Provisional Government, first in Debrecen, then in Budapest. These representatives could not assume the formal title of Minister or Ambassador because such an appointment called for reciprocity. For now the government was in any case duty-bound to channel all its communications to foreign states through the Allied Control Commission (ACC), the ultimate arbiter of the country's foreign policy. But the premier's office and the foreign ministry neverthe-less maintained more or less informal relations, especially with representa-tives of the western powers. The intent on both sides was to bypass the Allied Control Commission and withdraw certain sensitive questions from its scrutiny.

The first among the equals in this small, semidiplomatic corps was the soviet representative, Georgy Pushkin, officially "political adviser" to Mar-shal Voroshilov, who chaired the ACC but who for long periods of time was in Moscow, with his place on the commission being taken by a General Sviridov. But in contacts with government agencies Pushkin was the most conspicuous presence. He had gone to Molotov's school and had the hu-morless, inflexible personality of his boss. In a report to Washington the American representative wrote that his soviet colleague's eyes and expression methods contrary to Yalta principles) and that the receiving countries send nondiplomatic observers of their own.[5]

When Charles Bohlen, the chief expert in the State Department on East Europe, supported the proposal, instructions went out. Those to Schoenfeld, received on May 22, read: "In establishing your contacts with Hungarian authorities . . . you are authorized to say that in view of the large Hungarian colony in the United States and other Hungarian interests here, this Government would have no objection if the National Provisional Government of Hungary should send a representative to Washington." This step, the instructions emphasized, did not mean the resumption of diplomatic relations. The representative in the United States would nevertheless have access to the authorities and could represent Hungarian interests.[6]

Schoenfeld delivered the message to János Gyöngyösi on May 26 and drew the regretful reply that the Hungarian government had been given to understand by the soviet occupiers that "all contacts with foreign governments must be conducted through the ACC and have its approval."[7] Gyöngyösi subsequently wrote to Voroshilov asking for permission to comply with the American proposal; he had no reply.[8]

Possibly the marshal was awaiting the outcome of a broader initiative, from Stalin himself. The soviet dictator had written to Truman and to Churchill, informing them that as Rumania and Bulgaria "have proven in reality their readiness to cooperate with the United Nations," the soviet government deemed the time right to resume diplomatic relations with them. The USSR also contemplated taking up such relations with Finland and, "after a certain amount of time," with Hungary too.[9]

The singling out of Rumania and Bulgaria as deserving of immediate recognition and the mentioning of Finland and Hungary only as an afterthought was a maladroit move on Stalin's part, likely to raise suspicions in the west. It was precisely in Rumania and Bulgaria that political coercion and intimidation by the communist-controlled governments was most flagrant, even though London and Washington protested repeatedly. Stalin's plea must have had the purpose of endowing those two tightly controlled satellite states with a legitimacy they lacked in their unrecognized condition.

Perhaps had the soviets been willing to allow the western representatives on the ACCs greater scope, Britain and the United States would have been ready to trade off the recognition of unpalatable governments for being active players in the communist orbit. But their subordinate status remained a constant irritant and a practical guarantee that every one of Stalin's requests would be turned down. Nevertheless, Averell Harriman in Moscow advocated a positive response to Stalin's overture. He wrote to the

State Department that "we will find it difficult to get the Russians to any real tripartite basis for action in the Control Commission" and that the United States could do better "by handling as many questions as possible directly with the governments concerned."[10]

The State Department was unimpressed. It might have reacted differently had Stalin suggested the recognition of the Finnish and Hungarian governments first, to be followed in due course by the regimes in Bucharest and Sofia. It would indeed have been most awkward to take up relations with governments that were being severely criticized for excluding prowestern elements from their political life and for suppressing all opposition. There were also the economic questions, particularly weighty in the case of Hungary, which, by Stalin's plan, would have been a candidate for recognition only in the uncertain future.

Ever since the outbreak of World War II western properties in East Europe had been in great numbers either nationalized or outright confiscated. The Hungarian subsidiary of the Standard Oil of New Jersey, the Hungarian-American Oil Industries Company (MAORT), was a case in point. The company's properties, valued at $58 million, consisted of wells, pipes, warehouses, and other buildings and premises; they had been taken over by Hungary's wartime government in 1941, six months after the country entered the war. With the war now over, MAORT's status was uncertain. The State Department, at the request of Standard Oil in the United States, directed Schoenfeld to inquire. "Standard reports its fear that crude oil, refined products and petroleum equipment may be taken from MAORT and removed to Russia either in payment of reparations under armistice or otherwise."[11]

Schoenfeld reported back (on May 25) that MAORT properties were still in Hungarian hands but entirely under soviet control.[12] This state of affairs, not different from those obtaining in other soviet-bloc countries, again raised the question of whether recognition or nonrecognition of the governments involved would serve American purposes better. The State Department, supported by the president, took the position that nonrecognition would discourage American investment in those countries and enable the United States to make eventual recognition subject to the satisfaction of its pending claims. But the whole question of nationalized properties was closely connected to the nature of the government that prevailed in any given state, and in May the State Department was still "very satisfied with the representative character of the Hungarian government."[13]

By the end of the month, however, possibly because of the MAORT affair, or because of the concern over developments in other soviet satellite countries, the official view regarding Hungary changed. Another State Department memo argued that while on the surface the government did not appear to be totalitarian, communists held the key positions in every ministry and the State Department had information that moderate elements in the Provisional Government intended to resign because of their lack of influence over major decisions.[14]

The month of May, 1945 seems to have been a watershed in Truman's appraisal of soviet designs. He had never been a warm friend of the USSR and as senator had made some blatantly unflattering remarks about it. On becoming president, he realized the importance of dealing with the soviets in the interest of peace. But soviet conduct in East Europe, especially in Poland, so nettled him, as it had his predecessor, that he extracted Harry Hopkins, FDR's old go-between, from his sickbed and sent him to Moscow with instructions to impress on Stalin the importance of respecting the Declaration on Liberated Europe drafted at Yalta. While Hopkins was in Moscow, Truman formulated his reply to Stalin's earlier proposal for the recognition of governments in East Europe. He declared himself ready to grant such recognition only to the Finnish government, not to the other three mentioned in Stalin's note. "In Hungary, Rumania, and Bulgaria," he wrote, "I have been disturbed to find governments which do not accord to all democratic elements of the people the rights of free expression, and which in their system of administration are, in my opinion, neither representative of nor responsive to the will of the people."[15]

The message went to Hopkins first and he advised delaying its delivery until his discussions with Stalin over Poland had been completed, lest the issue of former enemy countries got "entangled with the Polish question in Stalin's mind and prejudice the chance of an acceptable agreement . . . regarding the Polish consultations." Truman agreed to the delay "until you consider the time appropriate."[16] When Stalin finally received the president's message, he took umbrage over Truman's readiness to take up diplomatic relations with Finland, which had not participated in Germany's defeat. "The public opinion of the Soviet Union and the entire soviet command would not understand if Romania and Bulgaria, whose armed forces participated actively in the defeat of Hitlerite Germany would be put in a worse position as compared to Finland," he wrote.[17]

On the English side, Prime Minister Churchill was of two minds. He had written to his political representative in Helsinki at the time he received Stalin's message that "we ourselves have been considering our future relations with Roumania, Bulgaria, Hungary and Finland, and we hope shortly to put comprehensive proposals to [Stalin] and the United States

Government." In a later telegram he deemed the time appropriate for concluding peace with the states in question. "It is not possible to restore formal diplomatic relations without first concluding peace, and no advantage is seen in entry into informal diplomatic relations on the Italian model."[18]

But coordinating positions with Washington proved problematic. Churchill took the characteristically European view that Americans did not understand the realities of the continent and that, as he wrote in a message to his ambassador in Washington, Lord Halifax, Americans "seem to think that, given the settlement of a few outstanding problems and the enunciation of general policy principles and desiderata Europe can safely be left to look after itself and that it will soon settle down to peaceful and orderly development."[19]

This of course was an entirely gratuitous view, unsupported by evidence. It was actually the British who took the position that direct and indirect soviet conquests in East Europe should be fatalistically accepted and that peace should be made without fretting over the types of governments that put their signatures to the treaties. Washington saw no advantage in such cavalier abandonment of nations to communist rule. True, where Poland, Rumania, and Bulgaria were concerned, it had more or less reconciled itself to the status quo. But in the case of Hungary it was determined to make a stand.

Chapter 7

The Economics of Poverty

The political consequences of the First World War had effectively destroyed the broad economic unity of the Danubian region. In the place of one large empire came five smaller states and the formerly unimpeded flow of goods ran into nationalistic obstacles. Austria was neutral and its international relations were on the whole amicable but Hungary, placed in the center of a circle of states that had all taken territory from it became the main obstacle to economic cooperation. Czechoslovakia, Rumania, and Yugoslavia combined in the Little Entente in opposition to Hungarian revisionism and Habsburg restoration. This hostile alignment itself prejudiced trade relations, but there were other problems as well. Rumania and Yugoslavia had heavily agrarian economies, with over three-fourths of the population engaged in agricultural pursuits. There was little they could offer to Hungary, or Hungary to them. Czechoslovakia alone had an industrial character, with only 28 percent of the population living off the land. But for seven years after Trianon Hungary refused to establish trade ties with a country that now included the entire northern part of the historic kingdom. Only in 1927 did more sober counsels prevail; a three-year agreement was signed for the exchange of Hungarian food products for Czechoslovak manufactured goods. By the time the agreement lapsed the Depression had struck, economic priorities shifted and bilateral trade came to an abrupt halt.[1]

During the early 1930s a number of alternative plans for the cooperation of the Danubian states were being discussed to allay the worst features of the economic crisis, but none resulted in agreement. The negotiations showed clearly that in an atmosphere of retributive nationalism trade relations suffered even when they would benefit all parties concerned. By the latter half of the 1930s Germany became the dominant economic power in East Europe, and it remained so during most of the war.

Now, in 1945, Germany lay defeated and the Soviet Union, which inherited its East European dominance, was impoverished, its infrastructure devastated. The soviet conception of foreign trade with states in its power sphere was disguised exploitation. This applied to Hungary and the situation was complicated by the fact that no asset or productive enterprise within the country was safe from soviet confiscations. Moscow's method of stripping an economy to its very foundations was amply demonstrated in occupied Germany. Beyond dismantling industrial equipment in their own zone, the soviets negotiated the right to dismantle and carry away machinery in the western zones as well, up to 25 percent of the total stock, on condition that they supplied food, oil, potash, and other commodities from the eastern occupation zone, though of somewhat less value than that of the goods they received. They carried out the first part of the agreement in their rapacious manner but failed to comply with the second on the argument that the machinery they were carting away from the western zones was war booty.

It was this putative claim to war booty that rendered property rights to a whole array of assets so dubious. In Hungary it made it practically impossible to prepare an inventory of national wealth; in addition it had the consequence that the British and Americans were holding back on returning the huge quantities of goods the Nazis had carried to Germany for fear they would end up in soviet hands.

Another example of how Hungary and Hungarians were victimized by the practice of war booty was the so-called CASBI directive issued by the Rumanian government. It was based on the most nebulous provision of the Hungarian armistice agreement. Article 8 bound the Provisional Government to disallow, without the authorization of the Allied Control Commission (ACC), "the exportation or expropriation of any form of property, including valuables and currency, belonging to Germany or her nationals, or to persons resident in German territories or territories occupied by Germany."[2]

It should be remembered that the armistice terms were drafted by the soviets and the Hungarians were able to obtain only minor changes. While on first reading Article 8 sounded fair enough given the enormity of plunder the Germans had inflicted on the Soviet Union, it was also an open-ended license to seize assets in the most diverse places on the most transparent grounds. When Bucharest issued the CASBI directive on April 5, 1945, it was initially only for "the collection of data," but it covered all manner of property belonging to Germans and Hungarians, or to those who had fled during 1944: land, houses, vineyards, tools and machinery, down to the most trivial items. The commissions carrying out the registration of goods chose the owners arbitrarily, often without any legal basis.[3] And inasmuch as the directive was based on provisions of the Hungarian armistice to which Rumania was not a party, it could only have benefited the USSR, with such ancillary gains as could be extracted going to Rumania.

Further confusion arose from the fact that the Rumanian armistice agreement invalidated the Second Vienna Award and ruled that Transylvania, "or the greater part of it," be returned to Rumania. This wording left the precise borderline in doubt. In applying the CASBI directive no one could be sure whether a given locality was subject to it. All this further aggravated an already festering ethnic conflict.

Yet when Hungary emerged from its commercial isolation, it made its first trade agreement with Rumania. It provided for Hungary delivering iron and seed grain in exchange for timber and salt.[4] A month later a similar agreement was concluded with Czechoslovakia; the latter undertook to ship coke and agricultural machinery against Hungarian shipments of oil-based products.[5] The monetary value of these contracts was so small, a total of 3.7 million peacetime *pengös* (approximately $800,000), that they were hardly more than of symbolic worth.

What Hungary desperately needed was trade with the west but that was not yet timely. It still had an outstanding debt of $580 million dating from the interwar years, owed in part to Britain and the United States and in part to their citizens or corporations. The continued unease of the western powers of what the soviets might do with shipments of great value in these postwar months was another vitiating factor. Thus for now the USSR was the only power that could rescue Hungary from its commercial, as well as diplomatic, isolation. Moscow addressed itself to this task with remarkable vigor, even though frictions abounded and the unruly behavior of soviet soldiery cast its shadow on the whole complex of relations.

János Gyöngyösi, who early in his tenure was accused of obsequiousness to the soviets,[6] by the summer of 1945 had recovered his courage and took every opportunity to register protests over Red Army misbehavior. When Georgy Pushkin, on July 9, brought Moscow's reply to the earlier request that Hungary be permitted to deport *Volksbundist* Swabians to the soviet occupation zone of Germany (a solution the reply deemed problematic, since a large part of eastern Germany had gone to Poland and the rest was in a hard-pressed economic condition), Gyöngyösi had a whole list of complaints to register with the soviet representative. Liberties taken by soviet personnel on the streets of Budapest were by now a standard lament, but there was also the specific instance when soviet military authorities demanded that a Budapest suburb provide quarters for 3,000 soviet soldiers within three hours.[7] Nor was this all.

The soviets had, by a treaty of June 29, 1945, annexed the Czechoslovak province of Ruthenia, claiming that the annexation was "[in] accordance with the desires displayed by the population of trans-Carpathian Ukraine."[8] Gyöngyösi had made a political speech in this connection, something he could scarcely avoid, as Ruthenia had been a province of the Hungarian kingdom for a thousand years. In the speech he pleaded for fair treatment of Ruthenians and Hungarians alike. He asked Pushkin to comment on the speech, drawing a guardedly approving reply. The envoy remarked however that he had detected a revisionist flavor in Gyöngyösi's comments. This was a grave charge, even when it was casually uttered; revisionism, of which Hungary was routinely accused by their neighbors, now also seemed to be directed against the Soviet Union. When Gyöngyösi asked for specifics, Pushkin said that there had been in the speech an undue emphasis on the rights of the Hungarian inhabitants. Did the foreign minister forget that Hungary had fought on Germany's side to the end? Gyöngyösi replied that the German alliance was Miklós Horthy's crime and not that of the Hungarian people. He added somewhat sharply that only because Hungary had renounced its historic claims it could not remain indifferent to the fate of millions of Hungarians outside its borders.[9]

Such spats aside (they deserve mention only because they were numerous and cumulative), on the highest level both sides made sincere efforts to establish firm good relations between the two countries. The soviets found it easy to pose as the only country concerned about the welfare of the defeated nation. On July 17 Minister of Commerce Ernö Gerö reported to the cabinet a soviet offer to conclude a trade agreement. It was to be substantial, certainly compared with the other such agreements, with a projected value of $25 million. The cabinet empowered Gerö to travel to Moscow and finalize the agreement.[10]

There was no more avid champion of close relations with the USSR than this man. Among communists he was known as a rigid dogmatist, who lacked Rákosi's subtlety of mind and broad culture, but to whom Rákosi deferred on occasion.[11] Yet when Gerö returned from Moscow with the completed agreement he professed to be full of misgivings. The soviets had pledged deliveries of articles essential for Hungary's economic reconstruction, such as coke, iron ore, cotton, copper and timber, but in exchange Hungary had to agree to conditions with long-term consequences: the establishment of joint ventures for the exploitation of the country's natural resources, the creation of new industries in partnership with the Soviet Union, a joint bank and navigational and air agreements. Gerö was a shrewd enough economist and soviet expert to understand what the term "joint" meant in soviet parlance. Through a whole series of fiscal and organizational expedients these enterprises would allow the soviets a dominant role and a disproportionate share of the profits. Another sticking point, a carry-over from the reparations agreement, was that the goods delivered by Hungary would be priced well below their current market value, whereas the accounting of soviet deliveries by and large reflected extant prices.

When the news of the commercial agreement was published, the political right and center reacted with extreme dismay. Industry and finance associations protested vigorously. Premier Béla Dálnoky-Miklós went so far as to tell Alvary Gascoigne that he would refuse to sign the agreement; a Smallholder complained to Gascoigne that the USSR was putting Hungary into a hopelessly dependent position and inquired whether His Majesty's Government (HMG) would not consider neutralizing this threat.[12] The British did not need much prodding. They first sent a note to Washington, complaining not only about the soviet-Hungarian trade treaty (its exact provisions were not yet available) but also about the agreements the soviets had concluded with Rumania and Bulgaria. Those, in the British view, were "contrary to the obligation of the Allied Powers *inter se* to abstain from negotiating arrangements with a common enemy." Referring to the Rumanian agreement in particular, the British embassy in Washington added: "A formal state of war with Roumania still exists and it follows from this that no one ally should unilaterally enter into arrangements which might prejudice the position of the other allies in eventual peace settlements. The Soviet-Rumanian agreement undoubtedly contravenes these principles and it is particularly improper in that it has been concluded with a government which owes its existence to active Russian intervention." As to the trade agreement between the Soviet Union and Hungary, "it appears . . . that it is very similar to the Roumanian and if this proves to be the case the above objections would apply with equal force."[13]

In Budapest even the communists took a less-than-cheery view of this supposedly evenhanded agreement. What somewhat blunted the indignation in every quarter though was that concern over the agreement's long-term negative effects seemed irrelevant in face of the immediate crisis produced by widespread starvation, the depradations of soviet troops, and the worsening inflation, all of which were really different sides of the same problem. The burdens of liberation had never been harder to bear. Rákosi wrote to Georgy Dimitrov that the distribution of meat in the country had ceased altogether. "The result is that the productivity of labor constantly decreases in spite of our greatest efforts [to the contrary]. A part of the laboring class leaves the workshops, goes to the provinces and become speculators in foodstuffs. . . . Our [political] enemies exploit the situation and hold us responsible for the difficulties."

Rákosi didn't go so far as to identify his party's coalition partner, the Smallholders, as the enemy (though from other things he wrote and said it was obvious that he thought of them as such) but referred to the "reaction" or to "fascists." Hungary's miseries were due to these elements—yet they could be held responsible only in absentia.

> The leading fascists have without exception fled. We have not captured a single minister or general of the fascist regime. The cases before the people's courts concern largely third-rate individuals, mainly such as have committed some crime against the Jews, and that is of course not the real thing. We have compiled a statistic and it turned out that more than 75 percent of those interned are industrial workers, so that we now begin their release.[14]

The spiritual fountainhead of the reaction, Cardinal József Mindszenty, would not relax his hostility to the regime, behind every act of which he perceived communist scheming. He was the most resolute opponent of the commercial agreement with the USSR and did not object merely to its specific clauses. When the cabinet gave its sanction to the agreement, he wrote another letter to the premier, begging him not to put it into effect. "It would deprive us of our economic and hence political independence. I call your attention to the immeasurable responsibility before God and history that this entails."[15]

Dálnoky-Miklós replied a few days later. The cabinet had indeed given its agreement to the trade pact, he wrote, but it was an agreement only in principle. The details would be negotiated separately in Budapest. "Thus your warnings are not yet timely."[16]

To the public, at home and abroad, Hungary presented the picture of a deeply divided political entity, no segment of which seemed to trust the others. Consequently the opposition to soviet influence was fragmented and ineffectual. In the view of the Hungarian Communist Party (MKP) the situation would not have been half as bad had it been able to introduce into the nation's political consciousness an ideological purity and endow the sacrifices with rhyme and reason. After all, as they saw it, the break with a deplorable past did not come cheap; the country had to atone for the crimes of the old regimes.

To the old guard this was seen as empty sloganeering. The communists were cloaking their drive for political power in trendy Marxist arguments while steadily strengthening their means of control. There had been nothing sinful about the past; Hungary had shaped its destiny in its own way as every other nation did. At home such arguments could be voiced only at great risk. The most stinging indictment of the political drift in Hungary came from abroad, from a former head of the Smallholders' Party, Tibor Eckhardt, now living in Washington. He wrote to his friends that he had long contemplated going home but had at last set the idea aside. "I haven't fought against the Nazis only to see the Bolsheviks come to power. . . . During the entire Nazi regime the hands of the Smallholders' Party remained clean. There was not a single Quisling among us."

He conceded that the armistice, hard as it was, had to be accepted, and that certain democratic reforms, such as land distribution were necessary to earn trust at home and abroad. But all those salutary measures had been politicized. "The land reform was not carried out . . . in a systematic manner but by local groups of a revolutionary character, recognized by the communists on the Trotsky model. These [groups] often employed summary, even terroristic, methods and brought on a class struggle; . . . worse, they undermined the authority of the government."

The same process was observable in the trials of the "war criminals." They had become political affairs and deprived the nation of any sense of justice. "Sooner or later justice of this sort degenerates into the persecution of the political opposition and creates an atmosphere of terror in which democracy suffocates."

The Smallholders' Party (KGP), the writer lamented, was constantly losing ground. The Social Democratic Party (SDP) had withdrawn from alliance with it and was marching with the MKP. "Don't they realize that after our party has been thrown onto the refuse heap, whose turn it will be next?"[17]

How far away that time was nobody could tell. Rákosi explained to the inner party circle that for now the Communist Party had to accept a coalition government because "even a theoretical discussion of our goal of a proletarian dictatorship would create an upheaval in the ranks of our coalition partners." Stalin, according to Rákosi, still thought time was on the party's side. He foresaw a communist takeover in Hungary in ten to fifteen years. That was when the inevitable crisis in the capitalist world would occur and it might well be followed by a war among the imperialists; this would be the opportune moment for Hungary and Czechoslovakia to enter the revolutionary road.[18]

Chapter 8

Emergence from Isolation

When the heads of government of the three great powers (France was not invited) met in the Berlin suburb of Potsdam on July 17, 1945 the principal item on the agenda was the treatment and governance of Germany, but questions relating to the East European soviet satellites also occasioned some heated debates. In the center stood American refusal (halfheartedly supported by Britain) to recognize the governments in Rumania and Bulgaria in their present composition. On the very first day of the conference the American delegation called for the reorganization of those governments to make them more representative by the inclusion of "all democratic elements," to be followed by "free and unfettered elections."[1] The soviet delegation rejected these demands with the same argument it would repeat in the future in diverse connections—that they amounted to an interference in those states' domestic affairs. To bolster their case the soviets cited the example of Greece, where just then British military force was being used against democratic elements.[2]

The debate continued in an ever more vehement tenor and in the end was postponed rather than resolved. Molotov promised that the elections would be open and the press free, but it was evident from the start that the western powers had no means to keep him, or the governments in question, to that promise. The United States had but one slender bargaining tool, the power to grant or withhold a soviet loan request for economic reconstruction (which had originally been for $4 billion but had since been reduced to $1 billion), but set against the enormity of the losses the soviets had suffered the loan amount was too small to be converted to political advantage. Repeated western demands for parity on the Allied Control Commissions of the defeated states was another futile agenda item. Given the resoluteness with which the British and Americans argued their cases in these foredoomed connections, it was surprising how readily they conceded to the USSR the right to seize German assets in countries that had been occupied by the *Wehrmacht*. While this provision had been included in the individual armistice agreements it had not until now received general sanction. The conference also authorized, in fact mandated, the expulsion of the *Volksdeutsche* from the East European countries Germany had occupied, although the American delegation maintained its opposition to indiscriminate deportations.

The license to expropriate German assets was one to which the soviets attached the greatest importance. Shortly after the Potsdam Conference Georgy Pushkin visited János Gyöngyösi and after a perfunctory demand for a list of war criminals in Hungary (so that it could be determined which among them were subject to international jurisdiction), he moved on to the question that interested him: he wanted to know what was the status of Hungarian assets in Rumania. Gyöngyösi was not slow in perceiving the purpose of the inquiry: some of those assets might turn out to be German assets in disguise. Gyöngyösi replied that no official records of such assets existed. A number of Hungarian-owned properties were stock companies and it could not be ascertained who owned the shares. The companies, he added, were reluctant to reveal their holdings for fear that "the allied powers" would lay claim to them. Pushkin sought to reassure him. The Soviet Union, he said, could claim only expressly German assets and it was in Hungary's interest to disclose these (presumably so that they would not be confused with Hungarian assets).

Gyöngyösi pursued his own line of argument. The great majority of Hungarian assets were in the American occupation zone of Germany, he said, and there was apprehension over bringing them back, "as the Soviet Union had taken a far-reaching and rigid position vis-à-vis us in the matter of war booty." The government feared therefore that the returned goods would be seized "by the ACC."

Pushkin indignantly protested against this assumption. But he shifted ground in the process. Whatever the Red Army had appropriated in Hungary, he said, it had earned through the shedding of blood. The Americans had not fought for the goods they had in their zone, Hungarian emigrés had taken them there. Hungary would now need the support of the ACC to get them back.[3]

It was impossible to know which of Pushkin's arguments and demands had come from Moscow and which were his own improvisations. Military commanders and diplomats often took broad liberties, as if standing on foreign soil had lifted from them the Kremlin's awesome authority. Some of their arbitrary decisions were allowed to stand; others were reversed, usually after they had been brought to the attention of a higher authority. In a number of towns the soviet command used the local branch of the political police to round up young men for "interrogation"; some of these were transferred to soviet custody for further investigation, then disappeared. Inquiring relatives received evasive answers, usually to the effect that the person in question was not being held by soviet authorities.[4] The actions were so haphazard, and varied so widely from one locality to another, that they could not possibly have been official policy.

Instances of arbitrariness were not limited to small transgressions. For example, on August 17 Béla Dálnoky-Miklós sent Gyöngyösi alarming reports of Ukrainian units of the Red Army crossing the border from recently acquired Ruthenia into Hungary, disarming the border guards, and advancing all the way to the Tisza River.[5] Gyöngyösi sent a note of protest to Klementi Voroshilov on August 21, adding that the news of the occupation had produced great consternation in the northeast of the country. "The Hungarian government and people," he went on, "have reconciled themselves to the joining of the Ukrininan people in sub-Carpathia to the USSR as a manifestation of a new order based on ethnic principles." But they had done so, he concluded, in the conviction that they had gained in a soviet Ruthenia "a good friend" and not a dangerous neighbor.

Voroshilov left the note unanswered but on August 28 a Hungarian officer on the scene reported that in the early hours of that day the last soviet soldiers had been withdrawn beyond the Trianon frontier. The officer suggested that in order to avoid further such incidents, border control should be exercised jointly by soviet and Hungarian guards. Until that was put in effect, Russian troop units should replace Ukrainian ones. He perceived a deep political motive behind the invasion, evident from the fact that the Ukrainian forces had distributed ballots and forced many inhabitants to vote in favor of their town being joined to the Ukrainian Republic.[6]

The practice of engineered plebiscites had become well-established in soviet imperialism. In the czarist period the will of the people had not been consulted and annexations rested on pure force. Stalin introduced a new device: consent by terror. Undoubtedly, had Voroshilov or those above him authorized a military takeover preliminary to a plebiscite, the public would have voted overwhelmingly in favor of being joined to the Ukrainian Republic of the USSR. This in spite of the fact that what reports arrived on the condition of Hungarians in Ruthenia were distressing. A good part of the manhood had been taken to concentration camps; only those who under duress declared themselves of Ukrainian or Czechoslovak nationality were exempted. Deportations to parts unknown were commonplace. Even women, especially teachers and civil servants, often fell victim to such summary action.[7] The fear that all this signaled further annexations grew stronger by the day.

The British and the Americans (especially the former) saw the conclusion of a quick peace as the best guarantee for these instances of soviet transgressions to end. Once the former Nazi satellites had regained their sovereignty and were recognized by other powers, infringements on their integrity would be violations of the international order. When the four foreign ministers met in London in September, their task was precisely the preparation of these treaties.

Molotov had an expedient formula: the peace treaties should be modeled on the armistice agreements and, except for minor alterations, should be identical to them. The British were on the whole agreeable, arguing only, in the case of Hungary, that Article 15 of the armistice agreement, calling for the liquidation of fascist organizations, need not be included in the peace treaty as it had largely been carried out. They stipulated however that "on the conclusion of the peace treaty all allied forces would be withdrawn from Hungary," except for such soviet troops as were needed "for the maintenance of their lines of communication . . . with the soviet zone of occupation in Austria."[8] This concession of course rendered the value of the withdrawal of "all" soviet forces nugatory, in the first place because it was ultimately Moscow's decision how many troops it needed for liaison, and in the second because even a token force had the potential of military intimidation. This was the position that the American delegation, headed by newly appointed Secretary of State James Byrnes took. One member later noted that, "It is hard to find in this project anything which meets our ideas of what a peace treaty should be. . . . In effect it reserves to the Soviet Union, and gives permanent character to, all the advantages of the surrender instruments, thus substituting . . . bilateral arrangements (economic topics) for the present method where at least some measure of joint Allied participation exists."[9]

Furthermore, he noted, "Presumably the Soviet troops would be withdrawn and military control terminated . . . [but] there is no definite provision for this, and the continuance of Soviet organs of control, for the fulfillment of reparations, obligations, and supervision of disarmament for example, may amount to an undercover control not much less effective than the open presence of troops." Worse still, there was nothing in Molotov's formula to guarantee the payment of debts owed by Hungary to Americans, "including those arising from Soviet removal of American property."[10] As to territorial questions, Molotov proposed omitting from the armistice text the qualifying phrase relating to Transylvania, "or the greater part thereof."

This was an instance in which, had the majority vote prevailed concerning property right to the land occupied by one power or another, the decision would have gone in Hungary's favor, even if the territory would not have been more than a strip of borderland. The position taken by the British Foreign Office was that there was no reason "to treat any of these three countries [Hungary, Rumania, and Bulgaria] more favourably than another and our general approach can thus be the same in each case." As to Transylvania, the Foreign Office was prepared to accept the restoration of the Trianon frontier. This concession was however followed by a circular reservation. "If the Russians wish to make minor adjustments in favour of Hungary, we should agree, but we should not support any major concessions to Hungary, as these could only be effected if they were agreeable to the U.S.S.R. and in that case the inevitable conclusion would be that Hungary was as much a satellite of the U.S.S.R. as Roumania."[11] Where the border with Czechoslovakia was concerned, it could be subject to minor modifications in Hungary's favor, "provided these were acceptable to Czechoslovakia."[12]

So much for the Foreign Office recommendations. Foreign Secretary Ernest Bevin took a mildly contrary position on the council. He said his only interest was in a "just and equitable frontier." French Foreign Minister Georges Bidault suggested that this might be achieved by an investigation on the spot as had been done in drawing the Italian-Yugoslav frontier.

Molotov would have none of that. He claimed (with some justification) that the ethnic groups in the frontier region were so intermingled that it was impossible to draw a line that would not leave a large number of people on the wrong side of the border. He reminded his colleagues that back in 1920 the chairman of the Trianon conference, Alexandre Millerand, had written to the head of the Hungarian delegation that: "The frontiers established for Hungary by the Trianon Peace Treaty are the result of painstaking study of ethnological conditions in Central Europe and of national aspirations." Molotov now argued that the British and American delegates had at that time accepted the dividing line between Hungary and Rumania without demur. As to the phrase "or the greater part thereof," referring to Transylvania, it had been included so that the hands of the peacemakers should not be tied in case new circumstances arose. "But nobody had suggested that new circumstances had arisen," Molotov concluded, suggesting that the Trianon frontier should be reestablished without alteration.

Bidault deferred but Byrnes did not. Already in 1919, he said, the United States had tried to secure a different border between Hungary and Rumania. Today it was possible, by a small shift in the line, to restore half a million Hungarians to their homeland. "In the area which he had in mind," Byrnes asserted, "there was a considerable Hungarian population, whose railway connections were almost entirely with Hungary and to put them into Roumania would contribute neither to their happiness nor to the happiness or prosperity of Roumania."[13]

The matter was tabled, but the outcome could not be in doubt. A year later, at another meeting of the Council of Foreign Ministers, Molotov would dispose of the issue bluntly by saying that all of Transylvania had to go to Rumania because Stalin had so decided in 1944.[14]

At this time, despite the intense nationalistic passions on either side, Hungarians were preoccupied with other questions than that of the location of their border with Rumania—in particular with national elections and the ratification of the trade agreement with the USSR. Both as they saw it would test the extent to which they controlled their political and economic affairs.

The MKP had raised the question of the elections in the spring, as soon as the last German soldiers were driven from the country. But the moderate parties were distrustful.[15] They feared that the communists would take maximum advantage of the presence of the Red Army and rig the results. Arthur Schoenfeld on June 18 proposed American-soviet negotiations on the subject. At the time the State Department did not want to prejudice the entire question of political freedom in East Europe, which was bound to come up at the Potsdam conference, by raising it prematurely. Should there be any sign, a department brief stated, that the elections were not going to be fair and open, the United States would demand international supervision.[16] But the political right in Hungary kept pressing Schoenfeld to intervene. He in turn warned his government that the political position of the Provisional Government was shifting to the left. "I respectfully recommend," he wrote, " . . . that we take early opportunity to make it clear to Soviet Govt that US cannot view with indifference institution of electoral law which will set political pattern of Hungarian State and determine future welfare of Hungarian people without thorough consultation and harmonious cooperation among Allies."[17]

Alvary Gascoigne reported to Churchill in a similar vein. He was of the opinion though that even international supervision would not ensure free elections because the communists were determined to come out winners. He suggested that elections be held after the conclusion of the peace treaty or, if that was not feasible, only after 1946.[18]

But rumors were soon current that Voroshilov, talking to the premier and the president of the National Assembly, had suggested October as a suitable time for the elections.[19] Gascoigne pointedly inquired of a foreign ministry official who visited him how democratic the elections were going to be. While he approved of the exclusion of fascistic factions from participation, he feared that the determination of what qualified as fascistic would be made by administrative rather than legal means.[20]

The MKP, gratified by a great increase in its membership, seemed confident that it could win at least a plurality without overstepping legal boundaries or relying on pressure from the Red Army, especially if the Social Democratic Party (SDP) could be persuaded to run on a joint ticket with it. The victory of such a ticket could later be hailed as a triumph of the left bloc. The Social Democrats, however, with a substantially larger membership than the communists, did not favor a fusion ticket—still, political pressures were incalculable.

Representatives of the western powers kept a close watch on internal developments for any sign that might point to undue communist influence. Schoenfeld thought he detected evidence of it. President Truman, in addressing the American people to report on the Potsdam conference, said among other things that the countries of East Europe, Hungary included, were not in the interest sphere of any great power. While the Hungarian press printed parts of the speech, it deleted the references to Hungary. The American envoy lodged a protest with Gyöngyösi but was told, as he reported it, that Gyöngyösi had issued a general directive to the press "not to publish statements which betrayed differences of view between allies." He added that the president's assurance fell in this "category of forbidden topics since passages in question were not published in Moscow."[21]

Rákosi, well aware that "the reaction" was cozying up to the western powers, made a surprise visit to Schoenfeld himself. Although his arguments were persuasive, they were not likely to elicit sympathy. The MKP, Rákosi argued, was Hungary's only hope for the future. No party could match its realism and vigor. It was regrettable that many Hungarians still looked to the United States to save them from trouble. Reactionary forces, Schoenfeld reported Rákosi as saying, "were laying ground work for distrust and disappointment in US by fostering impression that only salvation for Hungary lay in American help rather than in self help so that when it becomes apparent that US cannot meet exaggerated hopes corresponding anti-American feeling would develop here."

Schoenfeld further reported Rákosi as saying that although Hungary lacked the resources of a great power, it "had quite sufficient resources for modest national existence and should rather look to small states like Denmark and Norway as models of political and economic organization than aspire as visionary political leaders here commonly do to immediate adoption of American or British system which was the product of centuries of democratic experience lacking here." Rákosi incidentally impressed Schoenfeld as "forceful and highly intelligent with advantage of knowing his own mind . . . certainly . . . one of the more enlightened Hungarian public men."[22]

A few hours after Rákosi's visit the state secretary from the premier's office called on Schoenfeld and inquired, among other things, "if American troops will be here in October. . . . If only Russian troops should remain at election time non-Labour parties should wish postponement."[23] Within hours of this visit the government announced its decision for October elections.

The announcement occasioned a flurry of diplomatic activity. The SDP learned that the scheduled elections did not meet with the approval of His Majesty's Government (HMG) and that London might not even recognize the results. What particularly interested the British communicator was whether the two workers' parties would run on a joint ticket.[24]

The question was crucial because even a superficial fusion for election purposes would enable the MKP to dominate its sister party and, in case of victory, the entire political scene. When a member of the U.S. mission asked the head of the SDP, Árpád Szakasits, how large a vote a joint ticket would get, he drew an estimate of 60 to 65 percent. This was upsetting, not only to the Americans but to the British as well. As late as October 18 the head of the British mission saw fit to deliver to the foreign ministry a démarche stating that his government "will make it a matter of consideration whether it should receive the semi-official [Hungarian] mission sent to London" if elections were held on a joint ticket.[25]

As a dress rehearsal for the nationwide elections, Budapest held its own municipal tally on October 7. In this "red" city the communists and social democrats did run on a fusion ticket. The results were sobering. The ticket garnered 249,711 votes against 295,187 cast for the Smallholders. The other parties trailed far behind. The results dashed communist hopes that they would make a decent showing, let alone win, in national elections. In the provinces, despite their claims that they had been the moving force behind the land reform, they had only marginal support. When the politburo of the MKP met on the day after the municipal elections, Rákosi told his colleagues that the poor showing would have to be explained in the press as the effect of 25 years of Horthy and Szálasi propaganda that had permeated national consciousness and could not be undone within eight months.[26]

Soviet soldiery and the activists of the left bloc had behaved with exemplary restraint during the Budapest elections but no one could be sure that this would be the case when the whole nation went to vote. The most intriguing question was whether MKP-SDP cooperation was mere political expediency or whether there was an organic affinity that rendered differences between the two parties irrelevant. An American journalist sought out Szakasits to be enlightened on the matter. He wanted to know in particular whether relations between the two parties were good and why, if their programs were essentially identical, there was a need for two parties at all. He also wondered why the MKP, the more "leftist" of the two, pursued a more moderate policy than the SDP.

Szakasits explained that the SDP, over its 70-year-long career, had developed a mature membership, whereas the MKP, which had only recently appeared on the political scene, attracted many elements it could not absorb. Its new members were often inexperienced and aggressive. The two parties had decided on fusion for the municipal elections because most of the electioneering took place in factories and competitive campaigning would have hurt production, something the country could not afford. Szakasits emphatically added that the joint list had not been chosen on soviet command; in fact when the question was raised with a soviet officer he made it clear that it was an internal decision.

For all this, Szakasits did not think that a full-fledged fusion was timely.* As for the apparent moderation of the MKP, it was due to 25 years of intensive anti-Bolshevik propaganda which could be overcome only by a show of restraint.[27]

Drafting an electoral law had begun as far back as August. Rákosi had assured Gascoigne that it would definitely satisfy British opinion and that no party would be able to win an absolute majority.[28] The Smallholders did not take such a complacent view. One of their leaders, Ferenc Nagy, visited Gascoigne and asked him whether London had reconciled itself to Hungary falling into the soviet orbit. Gascoigne referred to Truman's August speech and expressed his own opinion by saying that the USSR intended to maintain its influence in Hungary only until reparations had been paid.[29]

---

*On October 12 the central leadership of the SDP decided against running on a joint ticket in the national elections.

On September 25 the head of the Smallholders' Party (KGP), Zoltán Tildy, was still telling Schoenfeld that his party was trying to decide whether to withdraw from the elections now or to wait until irregularities became evident.[30] Yet when the election law emerged, it satisfied even the most exacting observers. Schoenfeld had told Gyöngyösi on September 22 that if the elections did indeed allow free expression to all democratic forces, the United States would resume diplomatic relations with Hungary. This was far more than what Washington had offered so far, namely the willingness to receive a semiofficial representative. The decision apparently had been made in a hurry because Byrnes sent his instructions to Schoenfeld while attending the Council of Foreign Ministers meeting in London, without even communicating them to the State Department.[31] Three days later the Provisional Government pledged to Schoenfeld that it would meet the American desiderata for recognition.[32]

Now the USSR put in its own bid for Hungary's favor. As late as August 16 Gyöngyösi had still been telling Schoenfeld that Pushkin did not think Hungary should accept the American invitation to send a representative to Washington until it had sent one to Moscow.[33] On September 11 Dean Acheson still expressed himself in favor of a limited representation only. Then the one-upmanship began. On September 25 Hungary and the United States exchanged notes relating to the resumption of full diplomatic relations. On that day Voroshilov told General Key, the principal American officer on the ACC, that the USSR on its part was ready to enter into full diplomatic relations. "This action," Key wrote to the State Department, "was undoubtedly precipitated by the note which I delivered to Hungarian Foreign Minister September 22 expressing our readiness to establish diplomatic relations on certain conditions which has become common knowledge in political circles here."[34] Soviet control over Hungary had evidently not yet reached the stage at which Moscow would issue an outright veto over foreign policy decisions. All it wanted was to ensure that its own relations with Budapest should be formalized before the United States extended recognition.

In their somewhat undue haste the soviets involved themselves in contradictions. When they announced their readiness to exchange diplomats they cited Hungary's contribution to the ultimate defeat of Germany and the fulfillment of the armistice conditions—both dubious contentions considering that not a single Hungarian soldier actually fought against the Nazi forces and that soviet officials had repeatedly complained that Hungary was shirking its reparations obligations.

Voroshilov, perhaps trying to meet a deadline set in Moscow, personally delivered a letter to the Provisional Government announcing the resumption of relations; the president of the National Assembly signaled agreement on the same day. Four days later Moscow appointed Pushkin, by now well ensconced in Budapest, as its first official envoy. Budapest in turn nominated an old diplomat and public man, Gyula Szekfü, for the Moscow post. Thus, with formal peace still down a distant road, Hungary established a firm footing among "peace-loving" nations.

His Majesty's Government, which had argued all along for speedy recognition, now showed no hurry to follow the lead of its allies. The deputy representative in Budapest, William Carse-Mitchell, told a leading Social Democrat that he did not understand why Hungary wanted to hold elections at this early date when neither Czechoslovakia, Rumania nor Bulgaria had done so. Gascoigne went a step further. On October 18 he delivered to the foreign ministry a demarche suggesting that elections be postponed, though his other comments made it clear that it wasn't so much the elections HMG was objecting to as the proposed fusion of the workers' parties (the decision taken a week earlier against the fusion had not yet been made public.)[35]

American motives in taking up relations were not selfless either. According to one of Byrnes's aides, the purpose of recognizing the Budapest government was to "emphasize and give added validity to our refusal to do business with the present governments in Rumania and Bulgaria."[36] There was this consideration as well: if the United States wanted economic parity in Hungary with the USSR, it had to have a fully staffed legation and that was possible only if it extended diplomatic recognition. The recently concluded soviet-Hungarian treaty still awaited ratification and Washington was apprehensive that its purpose was to exclude western commerce from Moscow's sphere of influence. There could be no doubt that if the elections produced a leftist victory the new government would speedily ratify the trade agreement and perhaps make additional, more exclusive, ones. In speaking to Schoenfeld, Gyöngyösi tried to put a good face on the treaty, saying that its purpose was to facilitate expropriation of German-owned industries, but Schoenfeld was not satisfied. He feared that the pact would give the soviets a monopoly on Hungarian industry and questioned why, if its intent was to give the soviets access to German assets, the text did not make this clear.[37] (His apprehension was strengthened by a report making the diplomatic rounds, according to which Pushkin had told the president of the Hungarian National Bank that "Hungary should look to the Soviet Union for all her economic wants and need not entertain any ideas of western . . . ties.")[38]

Schoenfeld obviously overestimated the extent to which the USSR could satisfy Hungary's economic needs; at the same time, undeniably, one important goal of American postwar diplomacy was to enlarge commercial spheres and resist any attempt to exclude American business and industry from potential markets. This is not to suggest that the State Department was not deeply committed to the preservation of democracy and human rights; but the ideological and the commercial motives were so thoroughly intertwined that it was not easy to know in each case which of the two took prime of place. The institutions established for the purpose of restoring shattered economies—the World Bank, the International Monetary Fund, the United Nations Relief and Rehabilitation Administration (UNRRA)—were all in place, funded largely by American money, and while there were nobler motives behind each, as there would be behind that post–World War II American invention, foreign aid, the motive of enabling languishing economies to purchase American goods was never lost from sight. When the Americans—and later the English—insisted on free elections in Hungary, it was in good part because their political sympathies lay overwhelmingly with systems that practiced effective democracy; at the same time they knew that the economic door would be kept open only by the victory of the center and right-of-center parties, which favored close relations with the western powers. In Rumania and Bulgaria, where communist, or crypto-communist, governments were already firmly in place, their interests in economic relations were correspondingly cool.

Secretary of State Byrnes, perhaps afraid that the energetic Schoenfeld would overplay his hand, directed him not to press the Budapest government too hard to refuse ratification of the treaty with Moscow.[39] At the same time he instructed Averell Harriman to transmit to the soviet government a note expressing American interest in "mutually advantageous trade and other economic relations between Hungary and United Nations as soon as political conditions permit."[40] HMG fully concurred with this request; in fact it tried, unsuccessfully, to lay the matter of the soviet-Hungarian commercial agreement before the Security Council.

Schoenfeld, being, as every good envoy must be, a friend of the nation to which he was accredited, assured the State Department that the trade agreement with Moscow was generally profitable to Hungary. He later added however: "Creation of jointly owned company for exploitation of oil resources . . . will probably affect adversely American oil investments here," and that, furthermore, "companies for exploitation of coal and bauxite mining, aluminum manufacturing and electric power will result in Russian control of all important Hungarian national resources save possibly oil."[41]

This was what worried the State Department, not that the agreement might not be profitable to Hungary. It returned to the subject of soviet control of resources, and hence of markets, time and time again. Yet in this early period of relations the soviets acted with remarkable restraint. The election campaign was peaceful and there were no irregularities or instances of voter intimidation. Either Stalin wanted to gain a clear picture of the true alignment of political forces in Hungary or he really intended Hungary to remain independent, within indispensable limitations. Rákosi, depressed over the results of the Budapest elections, signed his name to an agreement that gave the Smallholders in advance a commanding majority in the National Assembly, as well as in the cabinet. His apathy was reported to be "induced by disappointment felt by soviet representatives in Rákosi's alleged misleading predictions regarding strength of communists in municipal election. . . . Voroshilov and Pushkin had not repudiated Rákosi but his attitude is almost the same as if he had been repudiated."[42]

Chapter 9

Bellicose Neighbors

The elections were postponed from a planned October date to November 4, mainly because of widespread concern over repressions and possible disorders. The Smallholders' Party (KGP) feared that the communists would use every means to prevent almost certain defeat. Voroshilov told Zoltán Tildy on October 17 that he had received reports of disturbances from many parts of the country and suggested that all the parties run on a joint list to lessen tensions. Negotiations were actually undertaken toward this end on the basis of Rákosi's proposal to concede in advance 51 percent of the seats in the National Assembly to the Smallholders. Tildy leaned toward the joint list, mainly to save face for Voroshilov and Pushkin, whose fall from grace, as Schoenfeld reported, "might lead to appointment of new Soviet representative here and much heavier hand . . . in handling Hungarian affairs."[1]

In the end the single-list proposal was abandoned; the elections passed without disturbance or any observable interference on the part of soviet soldiery or officialdom. Rákosi, when he recovered from the effects of his party's dismal showing, analyzed the results in a long letter to the editor of the Hungarian-American communist newspaper in New York:

> We have received 826,000 votes, 17 percent of the total. We have 70 representatives. Social Democrats, who hoped that as a separate party they will get substantially more votes than we, got only 20,000 more. In Parliament we are the third strongest party. We will have three votes in the National Council which acts in lieu of a president.

Numbers are misleading though. We are a homogeneous party, with no factions, no left or right wing. On the democratic left wing [of the political spectrum] is the National Peasant Party. It speaks for the poor peasants. They got 7 percent of the votes. Together with the Social Democrats the three leftist parties got a total of 42 percent. Considering [the effect] of 25 years of Horthy propaganda and that the election was secret, this is not a bad result. Consider also that at the time the elections were held we could not, in the capital and the industrial centers, honor the ration cards for 15 dkg of bread. Inflation has made paper salaries worthless. My own pay as Deputy Prime Minister is less than $4 [a month]. Reaction is in full swing against us. Mindszenty. . . . in a pastoral letter four days before the elections [warned believers] against voting for us. On election day the warning was read in place of a sermon. Peasant women [who have already voted] returned to polling places in tears asking to be allowed to invalidate their votes as they had just found out that they had committed a mortal sin [by voting communist]. Opportunistic parties used the presence of the Red Army against us in a demagogic fashion.

Half our votes were peasant votes, mainly from the trans-Tisza agrarian districts from socialist-Calvinist peasantry. 90% of the miners [voted communist]. . . .

Smallholders got 57 percent of the vote nationwide, 50 percent in Budapest. As it had no organization [in the capital] it is obvious that it got most of the reactionary votes. Only in the provinces was the actual majority of the votes [cast for] the Smallholders. Of these, many are fat kulaks.

One of their strongest weapons is chauvinistic agitation: in this they can rely on emotions whipped up in the Horthy era. Unfortunately Czechoslovak policies [against the Hungarian minority] is grist for their mills. Success of the Smallholders also had an effect in neighboring states where it is popular among our minorities, especially among those with fascistic leanings.[2]

The press in western democratic countries openly rejoiced, as much over the fact that the elections had been free and untrammeled as over the victory of the political right and center. The soviets were gracious in accepting the results and made a virtue of their restraint. Moscow radio said that "the vitality of Hungarian democracy has been so conclusively proven by the results that even western observers write about Hungary in a laudatory tone."[3]

On November 15 the KGP, with Tildy as premier, formed a coalition government that was speedily recognized by a number of states. (The country, as Rákosi observed in his letter, still did not have a president, only a collective body, the National Council, which carried out the functions of a president. The Protocol Department of the foreign ministry reminded the government that for the purpose of mutual diplomatic recognition this question had to be addressed, as no foreign envoy was officially accredited until he presented his credentials, signed by his head of state, to the head of a receiving state.)[4]

Washington, unconcerned with protocol, recognized the new government practically within the hour of its formation. There was now hope that profitable trade relations with the United States would follow. The Hungarian-soviet commercial treaty was still hanging fire. The National Assembly had not ratified it and the Provisional Government, on October 15, had proposed two significant amendments: that the agreement should not prejudice the right of Hungary to trade with other United Nations member states, and that the reference in the agreement to German property should, as per the Potsdam decisions, pertain only to such assets as could be put to productive use within Hungary.

But the KGP was still not satisfied. It feared the spirit of the agreement as much as the letter. János Gyöngyösi, in speaking to Schoenfeld, no longer had much praise for the agreement. He admitted that the soviets proposed it in the expectation that it would improve Hungary's chances at the peace conference; at the same time there was no reference in it to inter-allied cooperation for rebuilding the Hungarian economy; in fact, Schoenfeld noted, the draft had a "far-reaching unilateral character."[5]

Moscow denied this; Deputy Foreign Minister Andrey Vishinsky assured Harriman that the agreement was not intended to be exclusive and Harriman urged the State Department to accept this assurance.[6] The feeling was growing in American policy-making circles that further protests would only underline the helplessness of the western side and in the end provoke the soviets to stiffen their terms. The National Assembly apparently held the same view and when it met on December 20, representatives of the three major parties all spoke for ratification. A resolution in this sense duly emerged; however it was not unqualified. It recommended to the National Council "ratification of the soviet-Hungarian cooperative agreement and [took] note of the government's statement that this agreement in no way prevents the Hungarian state from concluding economic and commercial treaties with any other country."[7] On the same day the National Council gave its endorsement to the treaty. Despite the attached qualification, it was a triumph for the left.

The true intent of the pact with Hungary, as of those with other countries in the soviet sphere, was illustrated by the provision for joint-stock companies that was in many ways the principal feature of the agreement. Ownership was equally divided between Hungary and the USSR but with the initial capital investment originating almost exclusively from Hungary. The soviet contribution consisted of assets not originating from the USSR but of such as had fallen into soviet hands as reparations or confiscations. One historian notes that "the Soviet Union was able to obtain half ownership and full control of various enterprises in crucial sectors of the Hungarian economy . . . operating on Hungarian soil, through which it acquired extraordinary influence over the nation's economy. And Moscow achieved this result with virtually no obligation to export any capital whatsoever."[8]

Hungary had reason to wish that its relations with neighbor states had been as good as with the great powers. Actually, although the nationality problem continued to be a galling obstacle, Rumania and Yugoslavia were making serious efforts to mitigate it; the Rumanian government roundly denied that there was a nationality problem at all. The Hungarians in Transylvania were officially an integral element of the Rumanian nation and President Petru Groza missed no opportunity to emphasize that. The Czechoslovaks on the other hand not only admitted that there was a nationality problem but advertised their determination to solve it by the most radical means.

The Czechs and Slovaks were not entirely at one in this matter. There were liberal Czech politicians—President Eduard Beneš and Foreign Minister Jan Masaryk among them—who disapproved of the condemnation of an entire ethnic bloc, especially as the practice had acquired a racist stigma in Hitler's Europe. But these people were helpless in face of the hatreds the war had engendered. And once the great powers meeting at Potsdam sanctioned the expulsion of Germans from the Sudetenland, the practice was quickly extended to other offending minorities, quite irrespective of the degree of their guilt in the dissolution of Czechoslovakia.

Yet the Hungarian problem, as we have noted, was different from the German one, mainly because it dated so far back in history. It is true that Hungarians quite blatantly adopted all the slogans and attitudes of a master race, but it is also true that historic evidence seemed to bear out many of their contentions. Over the centuries the large landed estates in the northern uplands had been owned by Hungarians. The Slovak peasantry rarely rose above the conditions of serfdom. Commerce and industry were overwhelmingly in Hungarian (and Jewish) hands. Even after the establishment of Czechoslovakia in 1918 this distribution of wealth and power changed little and then largely for the benefit of the Czech element.

Only after World War II, in 1945, did the Slovaks at last have an opportunity to become dominant in their own country; their first agenda item was the purging of the Hungarian minority that once ruled them. Yet here, unlike in Rumania, the ethnic problem was defined by geography and, with different political realities, could have been solved by a modest cession of land. The northern mountains were heavily Slovak-populated; the southern flatlands north of the Danube were overwhelmingly Hungarian. The foreign ministry in Budapest prepared a memo on the question, explaining its position in historic terms.

> At the time of its occupation more than a thousand years ago, this territory was uncultivated and uninhabited. The Hungarians performed pioneer work, drained the marshes, and by their work. . . . they made the land fertile. For this reason the Hungarian nation claims that land, not under the always doubtful title of conquest, but under terms of the common law called by ancient Roman law *per specificationem*, meaning that the person who turns a worthless object into a valuable one shall become its legal owner, and of course also of the material used for this purpose. This nation which carried out all that work, and instituted governmental order in place of the former chaos, has the moral right to the respective territory.[9]

This argument, if carried to its conclusion, would call into question whether the Slovaks had any right to independent nationhood. Even in the Czechoslovak republic between the wars they were relegated to a second-rate status in practical governance. The central government and its organs were largely in Czech hands. The Czech-Slovak relationship had never been smooth. Days after Hitler had in effect placed the republic on the auction block at Munich, the Slovaks chose autonomy, and hyphenated their status in Czecho-Slovakia. Five months later they accepted from Hitler full independence and at once passed under the tutelage of the Third Reich. During the war they had a full-fledged Nazi regime, most of whose members after the war were tried for crimes against the nation, were found guilty and were executed. During the war the Prague government went into exile and from London presumed to speak for the entire nation. Putting the republic back together again after such antecedents was not a comfortable task.

The state, now deprived of its easternmost province, Ruthenia, still had to prove to the world that it was viable. Yet the old divisions continued. Most of the experienced statesmen, as well as businessmen and professionals, were Czech and the prospect of parity between them and the Slovaks was remote. For now however the much-publicized minority question kept the Czech-Slovak conflict out of sight. And although the position of the Germans and Hungarians was by no means analogous, once the expulsion of ethnically alien elements became national policy, it couldn't very well be applied to one group and not to another. And so the Czechs, in a lukewarm show of unity, joined the Slovak campaign against the Hungarian minority even though, as they soon realized, it was directed as much against the Hungarian state as against the ethnic Magyars in Slovakia.

Such a policy carried the seeds of dangerous consequences. The argument that the restored republic was punishing its despoilers could justify the deportations only so long. Once the argument lost its potency, the deportations would have to be stopped or Czechoslovakia would incur the odium of a racially intolerant nation. This realization might explain the inordinate hurry in which the "cleansing" process was being carried out. The Košice Program had laid down the principles, then decree followed decree.

On May 25, 1945 all state employees of Hungarian nationality were dismissed without any right of compensation; the pensions of those who had retired were cut off. The Hungarian language was banned from official intercourse, even in the purely Hungarian parts of Slovakia. No telegram or even letter in the Hungarian language was delivered by the postal service. A decree of June 22 confiscated all landed property of 50 hectares or more owned by Hungarians; another one of July 22 sequestered the cash and valuables Hungarians kept in banks. A still later decree made all Hungarians who had lost their citizenship liable to forced labor for reconstruction. The expulsion of thousands continued, while other thousands fled.

There was no international forum to which either the victims or the Budapest government could appeal; the Czechoslovaks carried out their actions by virtue of being a victor nation—although in all truth, apart from a feckless Slovak uprising as the Red Army already stood at the gates, the country had not struck a single blow for victory. Hungary's wartime credentials were however downright dismal. Rákosi kept putting his hopes in the better sense of his "comrades" in Prague. He wrote to Dimitrov: "We are hearing many complaints from Slovakia in connection with the comrades there employing indecent policies against Hungarians. Their defense is that the Slovak 'democratic' party is filled with [fascistic] Hlinka elements and that they thus have to exercise certain [discretion] in admitting Hungarian communists, for otherwise they cannot pose as a national party. The result is that . . . Hungarian communists feel that they are being oppressed and they turn to us for help." Rákosi begged Dimitrov to talk about the matter with Slovak comrades scheduled to arrive in Moscow for talks.[10]

In Hungary the coalition parties were genuinely perplexed by the fury and intolerance of the Slovaks, an intolerance that extended even to the Jews at a time when that harassed and decimated race enjoyed almost worldwide sympathy. The president of the Slovak National Council spoke from the heart of the nation when he said: "No difference can be made between Jews and non-Jews from the point of view of nationality. Those

Jews who by their language and way of thinking are to be counted as Hungarians or Germans must share the fate of Hungarians and Germans."[11] (Such talk, it should be noted, was bluster: very few Hungarian-speaking Jews in Slovakia suffered persecution or confiscation.) The situation was particularly delicate because Hungary itself had a sizable Slovak population that enjoyed full political and civil rights and was secure in its possessions; but if Slovakia continued its persecutions, reprisals would follow sooner or later.

As early as January, 1945 Molotov had suggested to a Czech Social Democrat population exchange as a solution.[12] In June, Gyöngyösi, speaking in his native town, which had a large Slovak population, made an emphatic point of Hungarian reluctance to initiate an exchange; the country would nevertheless accept this solution if it proved necessary, but only on the basis of an interstate agreement. If the Slovaks in Hungary chose repatriation, the government would place no obstacles in their way. They would enjoy the same rights the Czechoslovak government accorded to Hungarians.

This was an eminently fair position, but it did not bring with it international support for Hungary. The American opposition to arbitrary expulsions was as close as Hungary got to having a great power champion its case. General Key raised the question of expulsions from Czechoslovakia at a meeting of the Allied Control Commission (ACC), only to hear Voroshilov retort that he himself had received complaints but "in his [Voroshilov's] opinion they were based on rumors and did not have a factual basis." Key reported home that "Russians here taking no further action in the matter."[13]

The American ambassador in Prague, Laurence Steinhardt, took the side of the Czechs as a matter of course, writing home that "reports of persecutions and expulsions . . . have been grossly exaggerated and have been designed to operate as a spearhead [for Hungarians] to win the peace after having lost the war."[14]

In general the great powers took the comfortable position that such disputes should be resolved through direct negotiations between the interested parties. In November the Prague government invited Gyöngyösi for talks; he went early in December. By then the respective positions had been defined. For one thing it was clear that a population exchange, even if it was carried out in the fairest manner, would not bring about the ethnic purification the Slovaks desired; there were many more Magyars in Slovakia than there were Slovaks in Hungary and an even exchange (provided that all Slovaks in Hungary opted for resettlement) would still leave over half a million Hungarians in Slovakia. Budapest came up with a solution, though it was not one the Slovaks were likely to accept: Hungary was willing to take back its kin remaining in Czechoslovakia after the exchange, but only with the land they inhabited. In a purely geographic sense this was a reasonable prescription, especially commendable because about 75 percent of the ethnic Magyars were farmers and a resettlement without land would have uprooted them from the soil they had tilled for centuries.

Gyöngyösi's vis-à-vis in Prague was Vladimir Clementis; his extreme Slovak position in itself doomed the talks to failure. For four days he and Gyöngyösi wrangled. Clementis disclaimed any intention of judging Hungarians and Germans by the same measure but he insisted that both problems had to be resolved. He ruled out any border rectification, referring to Hungary's behavior during the war and its failure to live up to its armistice obligations (specifically the payment of reparations). He proposed a two-step process: first a population exchange, then the resettling of Magyars still left in Slovakia, except for those who had been allowed to retain their citizenship. The second phase would be unilateral, the movement of population in one direction only.

Gyöngyösi had few means to advance his own position. In tested diplomatic practice he started with issues of shared understanding, gradually moving to the more contested ones. He admitted that Hungary had been remiss in paying reparations but explained that the country's grave economic position made such payments impossible. He also reminded Clementis how long Hungary had endured in silence the harassment of its ethnic kin in Slovakia. He added that although his government opposed a population exchange, it would accept this plan in the interest of better relations. However he categorically rejected forcible resettlements. The Hungarians remaining in the republic after the population exchange had to be guaranteed full political and civil rights. While he respected the Czechoslovak wish to turn the country into a purely national state, Hungary would accept an unrequited transfer of population only if it was accompanied by land transfer.[15]

There was no ground for agreement. The matter would have to be referred to the great powers. The Council of Foreign Ministers was scheduled to meet in the spring to draft the text of the treaties that would be laid before the peace conference later in the year. They could not possibly disinterest themselves in the Hungarian-Czechoslovak dispute, even if they were inclined to do so.

Chapter 10

Preparation for Peace

Amid the crowding events of the first postwar year, relations with the Soviet Union remained the central concern of the Provisional Government. The signals from Moscow were mixed and inconsistent. Mátyás Rákosi had complained to Georgy Dimitrov as early as March 1945 that his party "received no advice to take a firm stand against reaction and against ministers who sabotage our work. Hence we miss many opportunities of [engaging in] the struggle that would increase our party's popularity and influence."[1] He seemed to assume that the great bulk of the nation had accepted the wisdom of Marxist teachings and wanted the Hungarian Communist Party (MKP) to lead it out of the wilderness of defeat. The political right and center on its part, while outlining a progressive agenda, sought through various expedients to preserve many features of the old regime, especially the sanctity of private property and of Christian morality. Moscow, aware of the broad appeal of the latter program and the political strength of its advocates, nevertheless made no effort to curb it. While unrelenting in matters of reparations and the seizure of German property, it showed only a most cursory interest in political developments. This stood in sharp contrast with its conduct in other East European states, where communist parties, generously supported by Moscow, maintained a tight control over political life. In Hungary the soviets seemed more intent on keeping the sanguine home-grown cadres, nostalgic for 1919, in check.[2]

A national conference of the MKP in May 1945 condemned the "left sectarian veterans of 1919"[3] and embraced coalition politics. In truth, coalition politics was something for which the party and its leadership were temperamentally unsuited. Militancy, an inseparable feature of the revolutionary ethos, and compromise, so essential in coalition politics, didn't go together. Yet Moscow insisted on the preservation of the coalition government and that was probably the only guideline the communists got.

The soviet leadership, as we have seen, endeavored from the start to present the Red Army as self-sacrificing liberators who would not try to impose political convictions on the Hungarian people or their government. There was no doubt a realization that the fortunes of the MKP depended to a large degree on the behavior of soviet soldiery. The high command was able to curb that behavior only spasmodically and tried to counteract it in other ways. In June 1945 it made a big food donation to the capital. In July it returned the coal mines it had seized to Hungarian control. In October the railroads were restored to government ownership. In August Klementi Voroshilov announced that all Hungarian prisoners of war (POWs) would, by order of the president of the Supreme Soviet, be released.[4] No other promise would more surely lift the spirit of the populace or create greater goodwill than the speedy return of loved ones. There is no reason to believe that the promise was made in less than good faith (though it might have been intended to influence the impending Budapest elections). In the next two months a few trainloads of prisoners came home, then the repatriation was stopped. Premier Béla Dálnoky-Miklós wrote Voroshilov a distressed letter. The enthusiasm created by the marshal's announcement had been bitterly disappointed, he complained. With the return of 2,847 prisoners by October 6 the process had been halted. Winter was approaching and "we are full of misgivings, especially as the return of prisoners, whom we sorely need for reconstruction, will be indefinitely delayed."

The premier referred to other concerns too: prisoners held in Hungary were kept in unsanitary, windowless quarters without adequate medical care. "I cannot conceal from you the fact that great unrest is caused in the public by reports that some 1800 soldiers, mainly officers,engineers and doctors, have been taken out of the country. It would greatly help if [this action] ceased and those who have been deported were returned." Prisoners passing through Rumania on their way home were reported to be subject to removal from the trains, at times by Rumanian Red Cross sisters, on the pretext that they required medical care—their fate was unknown.[5]

There was no clue to the precise soviet position on prisoner-of-war matters. Most likely the large-scale return of prisoners entailed logistical difficulties the soviets had not anticipated, or the political implications occurred to them only later;* in any case, it appears that the original offer had been made hastily and in the end caused only disappointment. Official relations were still uncertain. Georgy Pushkin continued doing what he had been doing before—acting as spokesman for Voroshilov—only with a different title now that he was an accredited diplomat. As to the Hungarian official presence in Moscow, the evidence is confusing. The newly named envoy, Gyula Szekfü, was in Moscow in early November; he sent a report on a conversation he had had with a Hungarian writer domiciled there who gave him the surprising advice that the Hungarian legation should direct the brunt of its efforts not toward the foreign ministry but toward artistic and intellectual organizations "which exercise a very strong and decisive influence on the policy of the USSR."†

In late December Szekfü was back in Budapest and had a visit from a counselor of the soviet legation who complained that the Hungarian mission in Moscow still was not properly staffed. It would be advisable, the counselor said, to tell the press that a staff would shortly be sent to Moscow and that the reason for the delay was that the legation building was being restored. (The statement issued to the press added that the restoration was a gift of the soviet government.)[6]

The rather unexpected courtesy that the Soviet Union and the United States were showing to Hungary raised hopes that their goodwill would carry over to the peace conference. But that conference would merely put a seal of approval on the decisions of the Council of Foreign Ministers and the council was so mired in controversies that the peace with Hungary was only a small item in a complex of unsettled problems.

On the German question the powers were far apart, the soviets proved difficult about the Italian treaty, and an East European settlement was complicated by the refusal of Britain and the United States to recognize the governments of Rumania and Bulgaria. During the second meeting of the council, in December 1945, the western powers relented on the last issue, on condition that the governments in Bucharest and Sofia

---

*When a year and a half later, a formal agreement for the repatriation of POWs was about to be concluded, Ernö Gerö wrote to Gyöngyösi that apart from the transportation problem there was a political one too: there were among the prisoners many who were "undesirable to Hungarian democracy," even dangerous elements, who had to be "separated" from the others.—New Hungarian Central Archives, Hungarian-Soviet Relations, Box 21, IV/438-3.

---

†"Despite the fact," the writer was quoted as saying, "that the form of government is a dictatorship, no enactment of major importance can be announced or carried out without previous consultation with these bodies."—New Hungarian Central Archives (NHCA), Hungarian-Soviet Relations, Box 5, IV-100-2.

broadened their bases with liberal elements. For Hungary this would mean an end to its privileged position as the only country in the soviet bloc that the great powers recognized.

Budapest was also chagrined that the powers showed so little interest in its conflict with Czechoslovakia. The British and Americans took the position that while they would not mediate, they would accept any agreement arrived at through bilateral negotiations. This position reflected either ignorance of historic realities—which was unlikely—or deference to the Soviet Union as the ultimate arbiter of disputes in East Europe. They could certainly not be blamed for it; some tough bargaining with the soviets still lay ahead on questions of far greater import and neither western power wanted to waste political capital on issues they could not substantially influence anyway.

If at least the Hungarian body politic could have shaped a coherent and consistent program to take to the peace conference! But the practice of bipartisan (or in this case multipartisan) foreign policy so dearly held in the west did not apply to East Europe. The Horthy regime had formed a national consensus by conducting an overheated revisionist campaign, fueled by public outrage over the Trianon treaty. But after a second lost war that consensus disintegrated. The Smallholders' Party (KGP) kept arguing for a joint platform. "Before all else," wrote one of their publicists, "we have to come to an agreement at home on what we want. The agreement . . . must obligate everybody; it is unthinkable that there should be two foreign policy positions."[7] But how to find common ground? The Smallholders would have no part of the communist argument that revisionism was a symptom of national smallmindedness, the very program that had carried Hungary into World War II and another defeat. As the communists saw it, greater tasks awaited the Hungarian people than pushing out their borders a few kilometers in all directions; the patriots on their part could not think of any task that was greater. They formulated, within the framework of the KGP, the principles on which the Hungarian peace proposals should be based:

1. Hungary was well on its way to becoming a genuine democracy
2. What had dragged Hungary into war was not the reactionism of the ruling classes but "their physical and spiritual distance" from the common people
3. Even in those ruling circles there had been many who fought against German influence
4. Hungary could not accept responsibility for the crimes of other nations (i.e. Germany and Italy)

5. Hungary should not keep silent about the persecution of her ethnic kin in neighbor states
6. Given the circumstances prevailing during the war, resistance to the Germans in Hungary had been no less creditable than in other nations, with the exception of Poland and Serbia.[8]

The last point was most popular with the public but it was also the least likely to get anywhere at the peace conference; it amounted to a plea of innocence in face of the fact that Hungary alone fought on the side of the Nazis to the end. The MKP on its part framed a policy of abject submission (knowing that this was what the soviets expected) but that did not have a shred of popular support behind it. The Social Democratic Party (SDP) attempted a balancing act. It examined the peace proposals without bias—and came away profoundly pessimistic. In a lengthy report the foreign policy committee of the party concluded that the victor powers would agree among themselves about the peace terms in advance and the Hungarian delegation would most likely not even be given a hearing. This amounted to saying that it was pointless to work out a peace plan at all—though this could not be said aloud. The report went on to predict that the western powers would not support Hungary's claims on Czechoslovakia, though they might be more receptive to claims on Rumania. Finally, any attempt to play the Soviet Union against the United States or Britain was doomed to failure.[9]

If the position against Czechoslovakia held any promise at all, it lay in the unreasonably harsh position the latter had adopted. Having gone on record as aiming at an ethnically homogeneous state, the Prague government left no room for compromise. It was doubtful though that international opinion would tolerate the expulsion of 600,000 people into an already overpopulated state that did not have enough farmland for its own cultivators. If Czechoslovakia stuck to its plan of expulsion it might be compelled to cede land, however modest in size, to Hungary. But that stage would not be reached until the two nations had agreed on a population-exchange formula and then the question of what to do with the Hungarians still left in Slovakia would arise. The Budapest government hoped that the western powers would not wait until this point was reached, precisely because the current Czechoslovak position went against all accepted notions of international justice. But on February 9, 1946 Arthur Schoenfeld delivered a note to Gyöngyösi's office, expressing regret that his government would not participate, as Hungary had requested, on an international commission to examine the respective claims of the parties.[10]

Meanwhile new negotiations had got under way in Prague; after four days of bargaining a tenuous agreement was reached, roughly along the

lines of the Clementis plan presented in December. Hungary agreed to allow members of the Slovak minority within its borders to opt for resettlement to Slovakia. It would admit a Czechoslovak commission to
propagandize the opportunities; the commission's members would have
complete freedom of movement and could use the press and the radio to get
the message across. It would have six weeks to carry out its campaign, after
which the Slovaks who opted for resettlement would have three months to
state that intention with the Czechoslovak commission. In exceptional cases
an additional month would be granted. The list of those to be exchanged
would then be presented to the Hungarian government, within a month of
the last deadline.

The agreement was heavily weighted in Czechoslovakia's favor. The
most glaring discrepancy between the rights of each side was that Czechoslovakia could select ethnic Hungarians to be exchanged while Hungary
could not discriminate among the Slovaks to be resettled.* A vaguely worded
article provided that the social and occupational composition of the exchange
groups should be approximately similar.

But the key problem of what to do with the Hungarians remaining in
Slovakia after the exchange remained unresolved. The Czechoslovaks
had the right to expel those guilty of war crimes by criteria set down by
the Slovak National Council, but the number of those was not to exceed
1,000. And that number was to be deducted from those sent over under
the exchange agreement.

In an annex the Prague government undertook to suspend further deportations, internal and external, as well as confiscation of property, until the
agreement had been signed and ratified. A further provision guaranteed
Hungarian civil servants and pensioners minimum means of livelihood. Both
sides reserved the right, should the agreement prove unsatisfactory in practice, to refer the entire question to the peace conference.[11]

The final agreement was signed in Budapest on February 27. It was
perhaps the best accord a defeated nation could hope for from a victor state,

---

*When Gyöngyösi brought the draft agreement before the parliamentary foreign policy
committee he expressed the fear that the Czechoslovaks would select leading elements, such as teachers, clergymen, and intellectuals, as well as well-to-do landowners,
then leave behind an amorphous mass. He explained that the removal of 1 to 2 percent of its membership could deprive any group of people of its identity.

Another worrisome provision was that most of the resettled would come from the
densely populated region just beyond the border and thus end Hungarian preponderance there; the Czechoslovak side even admitted that it wanted to be rid of that nearly
solid ethnic bloc.—New Hungarian Central Archives (NHCA), Parliamentary Foreign Policy Committee Minutes, Box 10, 4cc.

yet it showed how unsettled international relations were in this, the first year of peace. Hungary had not been at war with Czechoslovakia, in fact Hungary and Slovakia had technically been allies in the war against Russia. Had Slovakia come out of the war as a separate state it would have fallen under the same judgment as Hungary. But, benefiting from its position inside Czechoslovakia, it was able to pose as a victor state, or rather as the province of one.

Gyöngyösi had agreed to the treaty because time was pressing. The peace conference was due to meet in early summer and a preliminary Council of Foreign Ministers meeting was scheduled for May; if Hungary did not work out a bilateral treaty with Czechoslovakia as the great powers urged it to do, the Czechs could accuse it of a lack of cooperation and get the kind of treaty they wanted.

But there was still no agreement within the coalition in Hungary about the terms to take to Paris. The KGP, as we have seen, had formulated a position but it lacked concretion; Gyöngyösi was afraid that without a common platform the treaty he brought home would be assailed from all sides and lose its moral force. He explained to an interparty conference that no time could be lost as Hungary's hostile neighbors had already submitted formulas of their own to the allies.[12] Each party now went to work and produced a peace plan; the foreign ministry worked out one of its own. This latter one provided an extended historic overview that sought to explain how and why Hungary had drifted into the Nazi orbit. The process, according to the study, had started in 1919 when Hungary's "attempt at a democratic system" (the reference was to the ill-fated communist regime) was defeated by "Europe's reactionary forces in league with the rapacious imperialist policies of the successor states." It referred to the "dictatorship of Horthy" that established itself on a counterrevolutionary, anti-Semitic, and ultimately antilegitimist basis, opposed to the return of the Habsburgs to Hungary's vacant throne. Setting itself against all social reform, this regime was antagonistic to industrial workers as well as to the agricultural proletariat. To gain the support of the aristocracy, it perpetuated a semifeudal system. In time it reached out to big industry, a move that forced it to moderate its anti-Semitic stance. It also promoted a gentler and more civilized working-class movement in order to be able all the more effectively to assail the political left.

So the memo went on. It argued that Miklós Horthy's clique realized from the start that it could carry out its revisionist aspirations only in league with another dissatisfied power, either Germany or Italy. The ambivalent phase of Hungarian foreign policy began with Adolf Hitler's coming to power. But, in contrast with Hitler's Germany, Hungary tried to accomplish its irredentist goals peacefully, because it feared the shocks and economic changes a war would bring. This dilemma, combined with the public's hostility to Germany, explained why Hungary did not at once throw itself into the war against Russia as other satellites did. When Hungary finally did become belligerent, it sabotaged the German war effort and in this it persisted until the Germans occupied the country in 1944. At that point Hungary's sovereignty ended; the events that followed fell under a different evaluation. However, even after March 1944 Hungary's armed forces were not fully engaged against the soviet army until Ferenc Szálasi came to power.[13]

For all its obvious bias this study contained a great deal of truth, precisely because it juxtaposed the disreputable features of the Horthy regime with current restraint and peaceful disposition. Yet when this version was presented to a meeting of the coalition parties, the left bloc criticized it for taking the moral high ground. Gyöngyösi threatened to resign if he were not allowed to take this position to the peace conference. Other debates centered on whether to advance territorial claims against both Czechoslovakia and Rumania or against Rumania only. On none of these questions was unanimity achieved.

The MKP hedged on preparing a precise peace plan of its own. The political committee suggested that a delegation should first be sent to Moscow in order to clarify pending questions between Hungary and the USSR, and to solicit advice as to the position the Hungarian delegation should take to Paris. The hidden message was that the parties in the center and on the right were wasting their time formulating plans and marshaling supporting arguments; no plan had any value unless Stalin gave his consent to it first. As if to underscore this point the political committee, at its March 28 meeting, decided that the party should renew its contacts with the parties in Czechoslovakia and Rumania, because only in that way could reconciliation be achieved; the same resolution eschewed any expression of chauvinism or irredentism.[14]

It is hard to know how sincere this abnegation was. None of the communist parties in neighboring states displayed the same sacrificing spirit. In Czechoslovakia in fact the Communist Party was more militant on territorial questions than the other parties were. Most probably Rákosi and his colleagues knew that while the soviets might show flexibility in minor matters, they would not budge in territorial questions and so it was best not to raise them at all. This was the essential difference between the MKP and the KGP positions—while the latter believed that at the peace conference the will of the majority would prevail, the MKP was realistically aware that majority rule did not apply in questions relating to East Europe.

On the Council of Foreign Ministers the western powers had a three-to-one advantage and there perhaps some modification of the hard soviet position could be achieved. They did show willingness to work in that direction. At the September 1945 meeting of the foreign ministers, James Byrnes had advocated changing the Hungarian-Rumanian border in Hungary's favor, though for form's sake he linked the proposal to direct talks between the two nations. Bevin and Bidault supported the plan; Molotov opposed it, both the change of the frontier and the bilateral talks. The familiar arguments were tiresomely rehearsed. But the question must have been lurking in western minds of what would happen if the majority carried the vote and the soviets decided to sabotage it. Would there be a practical way to make the will of the majority prevail? The meeting broke up with the issue hanging fire. Subsequently the U.S. embassy in Bucharest pleaded with Washington not to make its advocacy of a border change public; it would practically ensure the victory in the coming elections of Petru Groza, who was a close friend of the soviets.[15] It recommended referring the question to the United Nations. Byrnes agreed, arguing that the border debates were "hypothetical"; the United States would at the right time express its sympathy for minorities living under foreign rule.[16] (When Schoenfeld learned of this position he branded it an abdication of responsibility, a turning away from a problem Washington deemed insoluble.[17])

In London the Foreign Office was lukewarm to the Hungarian claim; the farthest it would go was to recommend a measure of autonomy for Hungarians in Transylvania. It argued that without soviet concurrence Anglo-Saxon pressures to change the border would not get anywhere.[18]

Toward the Hungarian-Czechoslovak dispute the western powers took the same weightless, disengaged position, although they were by no means ready to concede that the republic fell into the soviet sphere. It was the memory of Munich more than anything else that inclined them to take the Czechoslovak view of things. Prague was aware of this and determined to reap maximum rewards. One newspaper suggested that Czechoslovakia should demand an enlargement of the Bratislava bridgehead at Hungary's expense on the argument that "the capital of Slovakia is within range of heavy artillery."[19] The newspaper even resurrected a claim dating back to the post–World War I treaty-making when the Slovaks, arguing that a large number of Slavs lived in the Hungarian-Austrian border region, demanded a corridor between Czechoslovakia and

Yugoslavia.*[20] Not only that, but once the border with Hungary was pushed beyond the Danube river, it might as well be extended eastward, pushing the border away from the Danube altogether.[21]

Clementis, the point man in Prague for negotiations with Hungary, did not venture to go so far. He did however demand an extension of the Bratislava bridgehead to embrace five Hungarian villages and he incorporated the demand in a brief prepared for the peace conference; he also requested allied supervision of Hungarian finances to ensure the payment of reparations. Failing that, Czechoslovakia should be given control over Hungary's natural resources and special rights for the use of rail and water routes.

The Rumanians too had formulated ambitious plans but Premier Groza pursued his case more diplomatically. There was no mention of expulsions and Bucharest had introduced several largely cultural measures to improve the position of Hungarians in Transylvania. In general Groza turned a friendly face in all directions. He constantly stressed the need for good relations with the Soviet Union; when Washington recognized his government on February 6, 1946, he spoke of great economic opportunities and solemnly pledged that Rumania would abstain from nationalizing its industries and natural resources. As to borders, he reasoned that any change in Hungary's favor would weaken his position and in consequence the future of democracy inRumania. He even found support among ethnic Hungarians. Members of the Hungarian National Union, frightened by the fate of their brethren in Czechoslovakia, renounced chauvinism, declared support for the Groza government, and even went so far as to endorse the Trianon borders.

In Hungary meanwhile recent election results infused the nationalistic elements with new confidence. The right wing of the KGP, which the communists referred to as the "clerical reaction," had always been outspoken about their aspirations, among which internally the restoration of the monarchy and opposition to land reform, and externally the recovery of lost provinces, had been the most prominent. The fact that over half the vote had gone to the Smallholders made it clear that these aspirations had wide popular support.

The left bloc however had a very different agenda. Throughout 1945 the MKP and the SDP had been agitating for a republic. In practical terms Hungary had been one, if an imperfect one, since 1919. It was true that Regent

---

*There were echoes of this position in Yugoslavia too. When the Czechoslovak minister of trade traveled to Belgrade to conclude a trade agreement, he entertained Josip Broz Tito at dinner. The latter, in his speech, remarked tendentiously that Yugoslavia and Czechoslovakia were only geographically separated. He concluded: "The Czechoslovak and Yugoslav republics must move toward the future hand-in-hand, because both countries have for centuries had a common enemy."—New Hungarian Central Archives (NHCA), Foreign Press Department, Yugoslavia, Box 6, 4/bc.

Horthy had never stood for election and had exercised many royal prerogatives, but he had firmly rebuffed attempts by the Habsburgs to return to Hungary and in his selection of ministers and other appointments he generally respected election results. Still, the Catholic clergy remained loyal to the former ruling house and throughout the Horthy era labored for its restoration. The alliance between throne and altar was very much alive. Cardinal József Mindszenty, after the war, established contact with Otto von Habsburg and later, after his arrest in 1949, admitted that he had hoped to form a federated Central European monarchy, a personal union between Austria and Hungary, and thought that the opportunity would come with the outbreak of World War III.[22] Such a confession, most likely coerced, must of course be treated with caution.

In December 1945 Mindszenty wrote to Zoltán Tildy demanding that the proposal for a republic be removed from the national agenda.[23] The Board of Bishops and other Catholic bodies went on record opposing the proclamation of the republic. But the great majority of the political establishment had more progressive notions. And so on February 1, 1946 the republic was proclaimed. Tildy became president and Ferenc Nagy, another KGP politician, premier. The new government at once became more assertive in foreign affairs, as if it was now speaking with the authority of an equal. Before the proclamation Tildy had still spoken cautiously when, as premier, he addressed the National Assembly. He had stressed the need for unfailing and expeditious compliance with the armistice terms and speedy trials for the war criminals; he credited Hungary's emergence from isolation to diplomatic recognition by the USSR and called for friendly relations with neighbor states.[24]

When Nagy now made his maiden speech to the assembly he sounded less accommodating. While acknowledging the important role the USSR had played in Hungary's liberation, he expressed his gratitude to the United States "for recognizing us" and his confidence that the Americans would without much delay return the Hungarian assets taken abroad. Furthermore, he stated that "The countries disposing of a surplus must be persuaded that an extension of credit to our honest Hungarian democracy does not involve risk . . . that the help extended to us is also a good investment." He did not say a word about compliance with the armistice terms or about swift punishment for war criminals. While expressing hope for good relations with neighbor states, he also called for a "reordering of the Danube Valley problem." He concluded: "The Hungarian question is not a local East European question that can be treated in isolation; it is an organic part of European and global political problems."[25]

Among the many items Nagy touched upon, the plea for loans was the most urgent. The destructions, the freebooting, and the pillage of the past two years had thrown the country into economic chaos. The currency was being swallowed up by a vertiginous inflation. The *pengö*, stable and respected in the interwar years, had been debased by immoderate spending, the introduction of occupation currency, and the depletion of marketable goods. Schoenfeld reported to the State Department on February 15, 1946: "Hungary's financial deterioration is now proceeding at a runaway pace. This week American dollar increased from 800,000 to over 1,800,000 pengö, prices more than doubled and currency circulation passed 2 million million. State expenditures in February are expected to pass 5 million million almost entirely by new[ly printed] currency." The next paragraph was even more ominous. "Time rapidly and inevitably approaching when Hungarian currency will cease entirely to be acceptable as medium of exchange. All economic activity will then stop except that which can be transacted on barter basis."[26]

No relief was in sight. The gold supply that once underpinned the currency was in Germany, unavailable to the Budapest government; industrial as well as agricultural production was so low that there was no prospect of increasing the purchasing power of the money in circulation. Yet, as the Americans understood only too well, leaving the process unchecked meant surrendering Hungary to soviet economic domination. Schoenfeld advocated UNRRA aid; he also suggested extension of credit to enable the government to purchase American surplus material left over from the war.[27] The central directory of UNRRA had already approved aid in the value of $4 million; now the Hungarian minister-designate in Washington, Aladár Szegedy-Maszák, went to the State Department to express his thanks but at the same time to confess that the aid covered only a fraction of actual needs. An earlier American suggestion that a three-power commission devise means of rehabilitating the Hungarian economy had foundered on soviet opposition.[28]

At the end of the interview Szegedy-Maszák called attention to the fact that the results of the Moscow foreign ministers' meeting (in December) were interpreted by various quarters as an effective division of Europe into spheres of influence. As he put it, "this was a development greatly feared by Hungary, since it was in a position geographically and otherwise where it would be mercilessly ground between the opposing weights of such alignments." However the warning missed its mark and the Minister was admonished against "uncritical acceptance of such press comments [which] should not be confused with official declarations."[29]

All in all these early months of 1946 were the Janus season of Hungarian diplomacy: a season of endings and beginnings, when the political establishment looked in two directions at once—toward the past, when the country still depended confidently on the west; and toward the future, which loomed ominously from the east. As yet the advocates of both orientations enjoyed freedom of expression, especially when they were in diplomatic posts in the free world. But few believed that this freedom would continue indefinitely. Szegedy-Maszák had warned the State Department official who heard his pleas for aid that "the Hungarian communists were endeavoring to make political capital out of the non-recovery of [the displaced] property and were using the situation to weaken pro-American sympathy in Hungary."[30] He was speaking the partial truth. The soviets, and the Hungarian communists, no doubt wanted to inherit an economically viable Hungary, not a bankrupt one; a premature sovietization might mean the loss of those displaced assets forever. At the same time it was out of character for the Kremlin to subordinate its political plans to economic considerations. If the economic questions did enter at all, it was only in a much larger context, defensible on Marxist principles.

The answer Szegedy-Maszák took away from his meeting with the State Department official was that the problem of the assets in Germany was "very complicated . . . currently under consideration." In diplomatic language this meant that the United States had adopted a wait-and-see attitude. It did not wish to tip its hand by either offering too much and seeing the returned goods fall into soviet hands or too little and watching Hungary collapse economically. The decision, as Hungarians saw it, would ultimately be made not in Washington but in Moscow. The fact was that all foreign policy items were in abeyance until the Soviet Union had made its position on each one clear. It was no longer enough to rely on Pushkin's Delphic pronouncements or to satisfy soviet wishes in small tactical matters; overall guidance was needed. It was for this purpose that the government appointed a delegation to travel to Moscow and try to read Stalin's mind as to the future he envisioned for Hungary.

Chapter 11

The Journey to Moscow

Ferenc Nagy had been premier for only two months when he was called upon to lead a top-level government delegation to the soviet capital. In his memoirs he left a colorful account of the nine-day visit, the pomp and circumstance that surrounded it, and his sessions with high-level soviet personages. Much of the time, it appears, was taken up with sight-seeing and social activities, though Nagy and János Gyöngyösi had some intensive negotiations with Molotov and Vladimir Dekanozov. What struck the premier was that the opulence of the receptions and banquets stood in such stark contrast to the general misery endured by the soviet people: there was an obscene surfeit of food, drink, and delicacies. It was also noteworthy that at a time when the USSR needed every able hand for physical reconstruction, the streets of Moscow were teeming with police and secret-service agents whose main task, apart from providing security, seemed to be to prevent the visitors from taking unscheduled side trips or deviating from the schedule in any way.

In formal talks the soviet leaders, especially Josef Stalin, were forthcoming in offering minor concessions. Stalin agreed that reparations deliveries be spread out from the six years stipulated in the armistice agreement to eight years and that delivery of finished textile goods under the commercial agreement be postponed by a year. But on other matters he was far less forthcoming. The dispute between Hungary and Czechoslovakia, a particularly sensitive issue, took up the bulk of the negotiations. Nagy and Gyöngyösi patiently explained that the Czechoslovaks were willing to grant citizenship to only those Hungarians who acknowledged their membership in the Slovak nation and these made up at most one-third of the total. The others were slated for expulsion. This was unacceptable to the Hungarian government and it was anxious to hear the soviet position. The plea was accompanied by references to the Atlantic and the United Nations Charters and policy statements on nationalities by Lenin and Stalin, urging cultural autonomy and political equality.

The reader will remember that in talking to the Czechoslovaks Stalin had repeatedly voiced understanding for their attempt to purge the nation of non-Slavic elements. But now he was facing the opposing side. According to Nagy's account he said, "I acknowledge that you are just in desiring equal citizenship rights for the Hungarian population in Czechoslovak territory and I state herewith that the Soviet Union will support such undertakings."[1] (This may or may not have been a verbatim quote; Gyöngyösi later reported it to the parliamentary foreign policy committee in a guarded version, saying that he hoped the soviets would support Hungary on this question at the peace conference.)[2]

On territorial questions Stalin held his cards very close to his chest. He had no argument of course with the Hungarian decision to renounce all land claims on Yugoslavia, but proved extremely reserved on the Transylvanian question. He had Molotov look up the text of the armistice agreement and after reading it acknowledged that the phrase "or the greater part thereof" gave Hungary the opportunity "to introduce [at the peace conference] the question of Transylvanian territory."[3] This was as far as he went, however.

Perhaps the members of the delegation could be forgiven for reading more into that pronouncement than Stalin had intended. None of them had any experience in foreign affairs and they were ready to hear "yes" when the answer was a "maybe" or even an outright "no." In any case, when reporting to the Hungarian public on their visit they tried to put the best interpretation on everything that transpired. It is also true though that Mátyás Rákosi had been vaguely hinting at secret information he possessed and allowed members of the delegation to believe that there was flexibility in Stalin's position.[4]

Even the economic concessions were self-serving because Stalin had by now realized that if the Hungarian economy continued to be overburdened with reparations obligations and deliveries under the commercial agreement it might break under the strain. By allowing more time for deliveries he gave the economy a much-needed breathing spell. He also agreed, with a wave of his hand, to forgive the $15 million Hungary owed for the railroads the soviets had built on its territory—no great favor considering that those railroads were now within Czechoslovakia and Rumania. He promised "to begin to return prisoners of war even before signing the peace," a promise Klementi Voroshilov had made half a year earlier with very meager results; Stalin qualified it by adding, "as transportation facilities and administrative procedures permit."[5]

Upon returning home Nagy reported to Arthur Schoenfeld and hinted at "positive results," especially, as Schoenfeld reported to the State Department, in that "he now had free hand to manage his Government."* Nagy then made his first overture for a similar visit to the west and expressed the hope that Washington would understand that it had been his first duty to establish personal relations with soviet leaders.[6]

When Nagy reported to parliament there sat in the galleries a British assemblage that had recently come to Hungary under the official title of "British All-Party Parliamentary Delegation" to investigate conditions. The delegation was impressed by the warmth and spontaneity of the atmosphere in parliament, and the relative freedom of expression. The delegation spent two weeks in the country, observing nearly every aspect of national life and in the end issuing a "unanimous report." It was a somber yet on the whole upbeat document. "Almost every acre of Hungarian soil has been fought over."

It continued:

> Material damage is enormous. It is stated that only 26 percent of all buildings in the Capital remained undamaged. . . . In the provinces most villages suffered damage. . . . Only one fire engine is left in Budapest. Population of the city reduced by almost half a million, that of the [entire] country by one million. . . .
>
> Despite the deplorable situation left by the war, there is a sense of purpose and political awareness shown by almost the entire population. . . . We are convinced that there is now established in Hungary the seed of a new democracy which, given encouragement and understanding, may finally establish itself along Western political lines. Certainly this appears to be the wish of almost all Hungarians.

The Communist Party, the report went on to say, was closely aligned with the occupation forces and "exerts an influence in excess of its backing

---

*In speaking to Schoenfeld, Gyöngyösi later referred to a speech Stalin made at a dinner given to the Hungarian visitors in which he "disavowed any intention on the part of the USSR to interfere in domestic affairs of Hungary citing Lenin's principle of self-determination from which USSR would not deviate. In same speech Stalin adverted to unprovoked Hungarian attack on USSR in 1941 but disclaimed any vindictive spirit as shown by fact that USSR had heeded Miklós Horthy's appeal for armistice."—Foreign Relations of the United States (FRUS), Washington, D.C., Government Printing Office, 1946, VI, 282, fn. 78.

in the country. It is evident, however, that they recognize that in the past they may have alienated public opinion by being almost more Russian than the Russians themselves and that they are now appealing more to national sentiment."

After references to the political police (whose alleged atrocities had been greatly exaggerated) and the food situation (in December of 1945 only 500 calories daily were officially available and mortality among children was as high as 48 percent) the report turned to Hungary's foreign relations. "Her future depends on the peace treaty she would get. Another Trianon and reconstruction will long be delayed." In closing, the report lamented the fact that the British political mission in Budapest was "out of touch with realities. It failed to establish close relations with leading political figures. They create the impression that Britain is not interested in Hungary's problems; this is all the more unfortunate because Hungary's public has strong pro-British leanings."[7]

The report was conspicuous for its omission of any reference to territorial questions, yet that continued to be the most live and hotly debated issue within the country and it was obvious that until the peace conference had made final dispositions it would not be laid to rest. It provided an ideological battleground between the political right and left. Every time the irredentist theme surfaced, a communist was sure to deliver a sermon on how that obsession was a leftover from a reactionary past. When Gyöngyösi reported on his Prague negotiations to the parliamentary foreign policy committee, a Smallholder member demanded that Hungary should under no condition accept the resettlement of Magyars, except if land cessions accompanied it. The MKP publicist József Révai at once raised his voice in protest. If Hungary took this position, he said, then territorial claims would have to be put forth in all directions (a specious argument as Rumania and Yugoslavia did not propose to expel their Hungarian minorities as Czechoslovakia did). "The result," Révai concluded, "will be that we will cause those states to pull together against our revisionism. . . . We cannot interpret the results of the Moscow visit to mean that we can demand as much land as we please. That would ignore the fact that we lost the war and would jeopardize our position."[8]

It was clear that if the most influential political party took this position, Hungary would not be able to speak with a single moral voice at the peace conference. The nationalist elements still hoped that even if the USSR was steadfast in its refusal to revise the borders, Britain and America, motivated by high principles, would override its objections. These hopes received some encouragement from the growing chasm between the Soviet Union and the western powers, which made it more than likely that the latter would stand their ground even on questions not of vital interest to them. Stalin on February 9 had made an "election speech" (he was contesting a seat on the Supreme Soviet) in which he struck a defiant and bellicose chord. Any accommodation between the capitalist states and the socialist world was impossible, he said, and friction between the two was on the increase. The Soviet Union had to prepare for the inevitable next war (into which it would surely be drawn as it had been into the contest between fascism and the western democracies) and had to ensure its readiness by at least two more Five Year Plans. A month later Winston Churchill, speaking in Fulton, Missouri, delivered a bristling answer, accusing the USSR of dividing Europe with an Iron Curtain. He added that his wartime experience with the soviets had taught him that they respected power alone, hence it behooved the Anglo-Saxon powers to preserve their military preponderance.

Shortly after the speech the deputy foreign ministers of the great powers met in London to prepare the ground for the Council of Foreign Ministers that was to meet in April in Paris. A preliminary meeting produced a faint ray of hope for Hungary. When the soviet delegate demanded that all of Transylvania be placed under Rumanian sovereignty, the American delegate proposed a slender amendment to the effect that Hungary and Rumania must discuss the matter between themselves with a view of decreasing the number of people who would live under foreign rule.[9] Even such an insubstantial qualification encouraged the die-hard champions of revision. It was by now perhaps more a matter of prestige than a concern over the fate of Hungarians in Transylvania. It never occurred to these people that, in view of the European tradition that a defeated state paid with a loss of territory, Hungary should feel itself fortunate that it was not asked to give up any. The communists kept harping on the theme of "ruinous chauvinism" instead of pointing out that not suffering further truncations was in itself a gain.

The advice that Hungary work out its differences with its neighbors through direct talks had by now lost all creative novelty. The Budapest government nevertheless felt duty-bound to make at least a gesture in this direction. Czechoslovakia was a hopeless case; the Prague negotiations had clearly shown that. Rumania was a better prospect. A top official of the foreign ministry, Consul Pál Sebestyén, was selected for the mission. He traveled to Bucharest and spoke first with the foreign minister then with the premier (on April 27, 1946). The former, Georghe Tatarescu, received him with great cordiality and verbosity. Sebestyén on his part, with no bargaining chips at his disposal, relied on platitudes. Hungary, he said, wanted to live in peace and friendship with its neighbors. No nation in the region had a

larger population of ethnic Hungarians than Rumania; his government therefore deemed it right to raise the question of border rectification before the
peace conference met. Tatarescu cut him off. No Rumanian statesman, he
said, could possibly make the question of Rumania's western border a subject of discussion. "For Rumania Transylvania, the cradle of the Rumanian
nation, is more sacred and precious than anything else."

Sebestyén reminded him that the actual borders of Transylvania were
not the same as those drawn at Trianon; the latter included territories that
could never be called the cradle of Rumanians. Tatarescu acknowledged
this but said that the historic and nonhistoric territories had become so tightly
fused that it was impossible to distinguish one from the other.

Petru Groza, when Sebestyén met with him in the evening, had different
arguments but his conclusions were the same. It would be improper, he
said, to discuss the question of Transylvania while the great powers were
having it under consideration. When Sebestyén reminded him that the great
powers would welcome a bilateral agreement between Hungary and Rumania, and that even the soviets encouraged it, Groza let the argument pass.
The breakup of Transylvanian unity would be a fatal mistake, he said. The
Vienna Award had shown that. Neither Hungary nor Rumania was the par-
ent of that province; it was known only by its own sons. There was no need
to talk about autonomy for the ethnic Hungarians, because Rumanian nationality law gave all minorities full freedom of expression.[10]

When the Council of Foreign Ministers met in Paris on April 25 to discuss the satellite peace treaties, none of the contentious points had yet been
ironed out. Haggling over the Italian treaty again consumed a great deal of
time and the question of Hungary's borders was not taken up until May 7.
On the evening of this day the shattering news reached Budapest that James
Byrnes had bowed to Molotov's demand that the borders revert to their
Trianon status. There was no more talk about letting Hungary and Rumania
settle the matter between them, though that subject was, because of Stalin's
unbending attitude in the matter, in any case passé. As to the Hungary-
Slovakia border, Prague had four weeks earlier asked the great powers to
authorize the extension of the Bratislava bridgehead, as well as the expulsion of Hungarians who would remain in Slovakia after the population exchange.[11] To grant the former demand would have gone against the
decision that the Trianon borders be restored and so the council decided
to treat it as a mere suggestion. The other demand was left entirely to
the consideration of the peace conference.

Disagreement arose over other matters though. Ernest Bevin demanded
that the soviets pull their military forces out of East Europe; he drew a sharp

rebuff from Molotov. The latter also clashed with Byrnes, over the question of Hungarian reparations. The secretary of state charged that the soviets were ruining the Hungarian economy by insisting on payments the destitute country was unable to pay; Molotov countered by charging that the displaced assets held in the American occupation zone in Germany were the main cause of the sorry state of Hungarian economy.[12]

Such questions, in the minds of most Hungarians, paled in comparison with the realization that there would be no "justice" after all where the borders were concerned. A deep gloom descended on the nation. Reports from envoys abroad did little to dispel the sense of discouragement. The minister in Paris, Pál Auer, wrote that Britain and the United States had genuine understanding for Hungary's position, but they would not oppose the USSR in East European affairs. The United States was determined to use its influence only in matters of prime interest to it and Hungary's borders did not fall into that category. Thus decisions regarding those borders had been taken in a matter of minutes.[13] From Moscow Gyula Szekfü reported that in view of the worsening relations between the USSR and the western powers the former was placing security concerns above all else and deemed Rumania more important than "the somewhat eccentrically located Hungary."[14]

Szegedy-Maszák in Washington wrote to Gyöngyösi in a similar sense. The conference had poisoned relations between east and west, he reported, and this probably meant an end to four-power cooperation. The United States now was determined not to retreat from Europe. The latest election results in various European countries showed that the appeal of communism had peaked. American policy was now "to outwait Russia."[15]

Where devastated East Europe was concerned, this policy meant that soviet political influence would be counterbalanced by America's preponderant economic strength. But it was precisely about this prospect that Schoenfeld worried. He wrote to Byrnes in Paris that American interests in Hungary were being steadily relegated to a subordinate position.[16] Soviet demands were so heavy that they left Hungary incapable of paying its other creditors. They even took priority over "Hungary's subsistence and rehabilitation requirements!"[17] He continued: "Economic changes imposed on Hungary by USSR in form of reparations . . . economic penetration . . . and restrictions on economic relations with countries outside Soviet sphere constitute burden largely responsible for rapid deterioration of Hungarian economy. . . ."[18]

All the while the MKP kept hammering away on the theme that the country's ultimate hope in all matters was soviet trust and support. Rákosi had raised some hopes before the foreign ministers' meeting when in a big speech in the provincial town of Békéscsaba he said in all seriousness that the one and only important factor for Hungary's expectations at the peace conference was that it had been able to gain the confidence of the Soviet Union.[19] Even the Paris decisions did not dampen the MKP's trust in the great eastern neighbor. Evidently Gyöngyösi was persuaded that Moscow might still relent on certain vital points; he asked Georgy Pushkin to forward to the soviet government a request that he be allowed to fly to Moscow to discuss items that would come up at the peace conference, notably the fate of Transylvania. Molotov sent back word that the decision to leave all of Transylvania with Rumania had been made at the initiative of the American secretary of state. When Gyöngyösi inquired of Schoenfeld about the truth of this statement the latter declared himself surprised. He had information, he said, that "initiative in Transylvania matter to detriment of Hungary's hopes had come from the Russians."[20]

Most bitter of all were the abandoned ethnic Hungarians in Transylvania. They faced a future of being separated from the fatherland for many years to come, or indeed forever. The foreign ministry received a passionate unsigned letter containing a plea not to abandon that ethnic minority. "For if you do," the writer asserted, "you will have to face the charge that every Hungarian [in the mother country] is a traitor, either in his person or as [a member of] the government that facilitates the endeavors for dissolution . . . of Hungarian nation, or who drives a wedge between Hungarians in the mother country and those in Transylvania, or if he does not act on the conviction that the Hungarian nation is one family, one community."

The government, the writer urged, should resign rather than sign the treaty so as to demonstrate its solidarity with its brethren across the border. An appendix called the future of Transylvania "one of the most important territorial problems in Central Europe." Because in question was a territory twice as large as Bohemia-Moravia and two-and-a-half times as large as Switzerland. It belonged to Hungary not only by historic right but as an organic part of the Carpathian basin. It was separated from Rumania proper "by a wall 1500-2000 meters tall. Beyond that you have [a] trackless, uninhabited strip." The broad gateways and the two major rivers all led into Hungary. "Economically it is a natural complement to the Great [Hungarian] Plains while the Rumanian plains have their own complement in the southern and eastern slopes of the Carpathians." Only Transylvania could provide eastern Hungary with stone and timber, coal and iron. And the Great Plains could best supply Transylvania with grain. Hungary could not solve its water problems on the Great Plains without controlling the headwaters of the rivers that all had their source in Transylvania: the Körös, the Maros, and the Szamos; only in this way could flood danger be avoided. Ethnically no group could call Transylvania its own. The number of Rumanians living there had been cited differently at different times. In the first four decades of the century it had varied between 2.8 million and 3.3 million. In the same period the number of Hungarians was between 1.7 million and 1.8 million. Other nationalities made up about 800,000 people. But only the Rumanians and Hungarians were densely settled in their respective regions.

As a whole, the author argued, Transylvania could not be called either Rumanian or Hungarian. As there was no ethnic solution, the province had to be divided, with close attention to economic and hydrographic features. Admittedly both countries would then have an odd shape and communications between the parts would be problematic, but there simply was no better solution.[21]

Arguments such as these had had some validity after World War I and they might have assured Hungary a more generous slice of Transylvania than it eventually received, but in 1919 the eastern part of Hungary was occupied by Rumanian forces and the country was fighting for its life under a communist government; military realities, although temporary, influenced the peacemakers. In 1945 and 1946, under more settled conditions, Hungary suffered from the weight of precedent; there was a border and whether it had been justly drawn no longer mattered.

It was Stalin who had said that the borders of Hungary would remain the same and it was he alone who could still alter them. An appeal to him was the last hope. A member of the armistice division of the ministry of agriculture* recalled many years later that the head of the division pleaded with the chief of the soviet delegation with whom he negotiated, arguing that Hungary had not joined the war of its own free will and could not on that account be deprived of areas inhabited by Hungarians. The Russian, General Mihailov, replied that the matter was of such importance that he had to consult Stalin; he immediately did so on the telephone. The reply he received was crushing. Stalin said that at Yalta the two other powers requested of him that he agree to the restoration of the Trianon frontiers. Stalin claimed he had a theoretical right to refuse, as the USSR had not been a party to the Trianon decisions. As Mihailov reported it, Stalin actually had felt that the ethnic and economic conditions in the Carpathian basin might have persuaded him to agree to the return of the heavily Magyar-populated regions to Hungary. But Horthy's rigid and treacherous attitude cost the

---

*Every ministry set up an armistice division to ensure compliance in matters falling within the purview of that ministry.

lives of hundreds of thousands of soviet soldiers. Thus Stalin decided not to oppose the request of the western leaders.

Hearing this, the Hungarians were so shocked that they were incapable of continuing the negotiations.[22] The account might be overdramatized but the argumentation was unmistakably Stalin's.

As Ferenc Nagy was preparing to lead a delegation to the western capitals he had at least this certainty, that territorial questions would not be a subject of discussion; every report he had confirmed that the western powers would not oppose soviet dispositions, and the soviet position was unmistakably clear.

Chapter 12

The Western Trip

B y Ferenc Nagy's own account the nation greeted the news of his planned western visit with enthusiasm.[1] After the meager results of the Moscow trip this was not surprising. The very fact that a Hungarian government delegation was welcome in the great western capitals was cause for optimism. There were solid practical reasons as well. The premier did not want his trip to Moscow to stand in isolation or to suggest that his government was unequivocally eastern-oriented. The fact that Hungarian goods still languished in the western occupation zones of Germany was another reason to seek direct contact with western statesmen. The opportunity to present Hungary's problems with neighbor states might also bear fruit at the peace conference. Undoubtedly too, Nagy and his colleagues were anxious to impress the British and the Americans with the depth of prowestern sentiment in Hungary and to make them understand that official utterances pointing in another direction were made under pressure from the political left. Finally, these officials, only recently raised to high positions from their provincial provenances, savored the prospect of seeing the world and meeting the highest personages in Washington and London.

As the communists with characteristic narrowmindedness saw it, the visit (in which, for larger political reasons, Rákosi also participated) was a phase of the Smallholders' counteroffensive against the growing ascendancy of the workers' parties. The battle was formally joined on May 22 when the

Smallholders' Party (KGP)demanded that control over the police, which was almost entirely in communist hands, be "equalized."[2] In the words of a communist historian, "The Smallholders' Party needed for this not danger-free action international support and was bringing tensions to a breaking point, confident that, bolstered by foreign successes, it could strengthen the coalition government of Ferenc Nagy."[3]

This was a gratuitous interpretation as Nagy had been contemplating a western trip long before the crisis over the distribution of police control erupted. But by the time the above-quoted assessment appeared, Nagy had been removed from office in disgrace and for communist purposes it was useful to assign a selfish motive to his every political action.

Promising signs were issuing from Washington. On April 24 the State Department advised Aladár Szegedy-Maszák that Budapest's request for a $10 million credit for the purchase of surplus property had been approved, and under extremely favorable conditions.[4] What a contrast this presented to soviet confiscations to which there seemed no end. Even the communist press made no attempt to assign nefarious motives to the American offer.

Nor was there opposition from the Hungarian Communist Party (MKP), whatever private reservations its leaders might have had about the western visit. The internal position of the party was still tenuous, it had little to show for its strident prosoviet stance, and the economy continued to go downhill. To veto the trip (without approval from Moscow) at a time when the entire Hungarian gold supply was in American hands, would have been a blatantly political move. Only the Social Democratic Party (SDP) was critical of the planned trip. János Gyöngyösi had long been a favorite target of that party; he was held responsible for a signally unsuccessful foreign policy and the failure of his ministry to compose party differences over the terms Hungary should propose at the peace conference. Gyöngyösi defended himself by claiming that Hungarian foreign policy operated within such narrow limits that its choices were illusory and on large questions it had to bow to the inevitable. The SDP was unimpressed; there is evidence it wanted to acquire the foreign portfolio for itself.[5] It was the only party with credible connections in both the east and the west; it had close ties with the British Labour Party while its left-bloc credentials kept it in good stead with the USSR. At the same time it was a singularly ineffective party in these early postwar years: it never defined its domestic or foreign agenda and made itself conspicuous through obstructionism, castigating Smallholders and communists alike.

SDPs opposition to the western trip had practical reasons behind it as well. The Hungarian envoy in London, István Bede, himself a Social Democrat, reported to his party's foreign policy bureau that the Foreign Office did not favor the visit since it had no chance of producing positive results.[6] The party thereupon sent a sharp memo to Zoltán Tildy, arguing that a time when relations among the great powers were so contentious was hardly auspicious for settling Hungary's problems. "We also believe," the memo stated, "that to take such a step when the Paris decisions [of May 7] have been so unfavorable to us, will unintentionally produce an impression that we seek the support of the western powers against the USSR; this is not only at variance with our fundamental policy conceptions but also amounts to utter naiveté and dilettantism."[7] Nagy finessed these objections by persuading Rákosi to be a member of the delegation. He told Rákosi he hoped to demonstrate that on vital questions there was total harmony between the party in power, on the one hand, and the party closest to the USSR, on the other.

But when Nagy informed the new chairman of the Allied Control Commission (ACC), General Vladimir Sviridov (who received him in the company of Georgy Pushkin), of the planned trip, Sviridov was sour. Nagy wrote later: "Pushkin [in particular] gave me the distinct impression that he would have advised us to stay home but, after the Paris decisions on May 7, found it wiser to avoid any attitude that might show that the soviet government intended to prevent Hungary from further asserting herself." Pushkin warned Gyöngyösi though against trading away Hungary's independence for a loan. Gyöngyösi replied that without a loan the economy could not be stabilized and reparations could not be paid. Pushkin replied, "Then you had better get a loan."[8]

Nagy in his memoir glosses over British reluctance to receive the Hungarian delegation. He writes only that "The British Minister told me that, while his government could not entertain us, it would be glad to receive a high-level Hungarian delegation."[9] In truth, the matter was much more complicated than that.

Alvary Gascoigne first advised the Foreign Office on April 26 of Nagy's intention to visit London. Ernest Bevin, in consultation with Foreign Office officials, decided early in May to honor the request; however it was not until May 17 that the decision was communicated to Budapest. (In a letter to Gascoigne the Foreign Office wrote: "It is to be hoped that such a visit would encourage him [Nagy] to stand up for the independence of his party in face of Communist and Soviet pressure and that it would also prove to him that we have [by] no means disinterested ourselves in the fate of Hungary."[10]

But soon Whitehall had a change of heart. On May 27 its chargé in Budapest, William Carse-Mitchell, told Nagy that because of his many other obligations Bevin would not be able to receive him and added that in view of the tense political situation in Hungary, it might not be wise for Nagy to leave the country, though he would at some later date be welcome in Great Britain.[11]

Evidence suggests that the Foreign Office was not so much impolite as uncertain as to how to handle the whole affair. It sought to avoid the impression that His Majesty's Government (HMG) was in league with the Smallholders in their "counterattack" against the left bloc. It also vaguely suspected that the proposed visit was a communist plot based on the expectation that an unproductive junket would compromise the Smallholders in the public eye.[12] Still another factor was London's reluctance to get involved in the Hungarian-Rumanian dispute, in which it had already taken Rumania's side as more important to its East European interests.[13]

The negative word from London was a shock to Nagy and his colleagues. Nagy himself was determined to make the western trip before the foreign ministers met for their last pourparler preparatory to the peace conference. One possible way to soften British opposition was to reverse the itinerary and visit Washington first. The finance minister, Ferenc Gordon, first floated this idea to Arthur Schoenfeld. He explained that he was visiting the envoy at the request of President Tildy, who was deeply concerned over the British attitude, especially as "recent international developments in Hungary, taken together with Communist victory in Czechoslovak national elections* would make any rebuff on the part of the western powers subject to interpretation here that Smallholders Party in Hungary and its representatives in Hungarian govt were discredited thus impairing their prestige and perhaps leading to fall of present govt." Tildy wanted to know if it might be possible for the delegation to go to Washington first, then stop off in London on the way back, at a point in time when the reasons that at present prompted HMG to ask for a postponement had passed.[14] Later in the day Nagy himself visited Schoenfeld and expressed the hope that "no such intimation as made by Mr. Bevin would be forthcoming from Washington.[15]

Schoenfeld assured Nagy that Bevin's dilatory attitude concealed no political motives but stated that Washington might give an answer similar to his as it received so many official visits that it was not always possible to arrange one at notice.[16]

London meanwhile had had another change of heart. It was possibly brought about by a communication from its ambassador in Moscow saying that in view of the fact that the Hungarian government delegation had been

---

*In parliamentary elections held on May 24 the Czechoslovak Communist Party achieved a plurality of 38 percent of the vote, which made it the largest party in the republic.

received graciously in Moscow, it would be unfortunate if its reception in London were poor in contrast, or if the London visit were indefinitely postponed.[17] In any case, on June 3 Carse-Mitchell received instructions to tell Nagy that Prime Minister Clement Attlee and Foreign Secretary Bevin would be glad to receive him and his delegation. Meanwhile Secretary of State, James Byrnes, had called Schoenfeld on June 3 to say that his government was prepared to receive the Hungarian party and would even cover the expenses of the trip.[18]

Nagy took care, before he left Budapest, to work out an agreement, though a fragile one, with the left bloc on the question of police control. This agreement was brokered by Pushkin. The Russian made it clear to Nagy that his government would not tolerate the dissolution of the coalition and urged Rákosi to make concessions. Rákosi, although he had pledged to Georgy Dimitrov that he would never yield on the question of police control, made a partial withdrawal. With the support of the SDP, he agreed to relinquish some 100 police positions held by communists. He also consented to the holding of early elections in the provinces; the abolition of the hated political police, the ÁVO, outside the capital; and the transfer to the Smallholders of government positions recently vacated by communists.

Four days before the departure of the delegation Gyöngyösi had a warning from Pushkin not to fail to remind western statesmen that Hungary was occupied by the Red Army and surrounded by Slavic states.[19] Rákosi's inclusion in the western visit was almost certainly not a result of soviet pressure; nevertheless his presence was a partial guarantee to Moscow that the KGP members would not take exaggeratedly prowestern positions; he would also prove a valuable source of information.

General George Weems, successor to General William Key as head of the American delegation on the ACC, placed his personal airplane at the disposal of the party; the ministers flew to Paris and thence proceeded by commercial aircraft to New York. After an unnervingly bumpy flight they arrived in Washington to a modest reception; the party was taken to Blair House, where it was housed during its stay in Washington.

Schoenfeld had advised the State Department of issues the Hungarians were likely to raise and they were familiar ones: pleas for economic aid and for support in effecting boundary changes. The envoy did not think Washington should oppose border rectification at Rumania's expense, especially as the communists in Hungary were passing the word that it was Byrnes who had proposed to keep the Trianon status quo. The slightest change in Hungary's favor, Schoenfeld wrote, would greatly strengthen the Nagy government's position.[20]

On the American side there was some apprehension about Rákosi's presence on the delegation; the State Department feared that as the only English-speaking member he would seek to monopolize the talks. The fear proved unfounded; Rákosi's role was largely confined to public relations. He spoke on the radio in English, gave interviews to the press, and was a visible presence on social occasions; only once did he participate in official talks and then said little.

On questions of the greatest import, concerning the chaotic economic conditions in Hungary and the maltreatment of the Magyar minorities, Rákosi's absence was of no consequence; he could only have echoed the arguments of his colleagues. In other matters Gyöngyösi had the opportunity to present the KGP position to his American listeners. He frankly admitted that political realities in Hungary had been shaped by the Red Army. A State Department official who spoke Hungarian recorded the foreign minister as saying that Hungarians were "overwhelmingly prowestern in sympathies." People of such sentiment, the official's minutes went on to state, made up about 60 percent of the population, including the independent-minded peasantry, and were anxious to preserve the middle-class style of living in Hungary. In this desire they were opposed by a minority composed mainly of industrial workers who were left-leaning and not interested in the continuation of a Hungary dominated by the middle class. It was obvious, Gyöngyösi added, "that the Left cannot obtain a parliamentary majority through constitutional means. . . . Nevertheless, the dynamic nature of the labor movement and the support that they are in a position to receive from the army of occupation might give them the upper hand in a political contest between the two opposing factions."[21]

John D. Hickerson, the State Department's Director of European affairs, neutralized Gyöngyösi's more pointed comments by referring to the common victory earned jointly by the USSR and the western powers in the war, a victory that led to the occupation of Hungary by the Red Army. He was noncommittal about other questions as well. This pattern of reserve continued through all the other contacts. President Truman received the party on June 13 but this was largely a ceremonial visit. A congressional group that entertained the visiting Hungarians at lunch was friendly but offered little support for border changes or for a loan from the Export-Import Bank, which had not deemed Hungary a credit-worthy applicant.[22] Nevertheless on June 13, the fourth day of the delegation's stay in Washington, Nagy was given good news: the Hungarian gold guarded by Americans in Germany would be returned.

In previous conversations with Byrnes, Nagy had been able to get a promise only for a partial return of the gold supply, the rest to be retained as a guarantee for the indemnification of American citizens for wartime losses. A day later that condition was dropped. Dean Acheson also assured Nagy that "American commanders in Germany and Austria are being instructed to proceed with restitution to Hungarian Government of identifiable displaced property removed under duress from Hungary subsequent to Jan 20, 1945 [the day of the armistice]."[23] He insisted though that Hungary had to agree to an aviation pact for which Washington had been pressing for some time. As to a credit line, the United States was ready to increase the current $10 million limit to $15 million.[24]

The delegation next flew to Tennessee to inspect the Tennessee Valley Authority (TVA) dam; Nagy in his memoirs comments on the breathtaking impression the view made on everybody in the party, with the exception of Rákosi who never missed a chance to make deprecating remarks about everything they saw, contrasting it to the magnificence of soviet achievements.[25]

In New York Gyöngyösi had a visit from a representative of the JOINT Distribution Committee who asked him about rumors of a new anti-Semitic wave in Hungary. Gyöngyösi challenged the assertion and said that the great masses of Hungarians were sympathetic to Jews and had been shocked by the deportations; if there was some anti-Semitism at present it was due to the fact that the JOINT, with large amounts of money at its disposal, had purchased stocks of goods in Hungary that were available only to Jews; the press had given this fact wide propaganda and it had produced resentment.[26]

The party next flew to London to a chilly reception. The Foreign Office official who acted as a guide made some unkind comments about the anomaly of Hungary, a defeated state, expecting territorial concessions from Czechoslovakia. Attlee, when he received the delegation, was also aloof and distant. Bevin had gone to Paris and the Hungarian party proceeded there next. Judging from Gyöngyösi's subsequent account to the parliamentary foreign policy committee, Paris was in many ways the most eventful stop, due largely to the fact that the peace conference was about to open and many important personages had foregathered there. Gyöngyösi had a meeting with Bevin, who informed him that the year before, in London, the Council of Foreign Ministers had agreed with Byrnes's suggestion that the Hungarian-Rumanian border be altered and it was a mystery to him how the proposition in the end came to naught. (Byrnes, when Gyöngyösi saw him, had an explanation: two American officers had gone to the border region and reported that the ethnic groups were so thoroughly intermingled that it was impossible to draw a line.) Bevin hinted though that as the memory of the war receded and passions cooled, especially after the Rumanian elections, it might be possible to draw a satisfactory borderline through a bilateral agreement.

On the Czechoslovak problem the foreign secretary was more encouraging; he expressed his opposition to a mass resettlement of the Hungarian population. At the same time he did not favor inclusion in the peace treaty of minority protection articles. The experience of the interwar period showed, he said, that such undertakings meant very little. Trust had to be placed in the United Nations Charter, which had an application going far beyond ethnic questions.[27]

Meanwhile Nagy saw Molotov, who turned toward him his most rigid, unyielding side. There was no possibility, he said, that the USSR would agree to altering the Hungarian-Rumanian border. He reiterated the claim that the initiative for leaving the border unchanged had come from Byrnes; the Soviet Union had agreed because it was in consonance with the armistice agreement.[28]

John Hickerson, whom Gyöngyösi had met in Washington, was in Paris at this time and the two had another talk. The American expressed doubt that the restitution of assets would bring about fiscal stabilization. He asked what percentage of the Hungarian budget went for the payment of reparations. (This to American minds was the truly important question.) Gyöngyösi could not furnish an exact figure but estimated that it was between 30 and 40 percent. Hickerson saw in that proof of his contention; if that much of a nation's budget went for unproductive uses, stabilization of the currency was out of the question. He inquired if Hungary had asked for an alleviation of the burden. Gyöngyösi replied in the affirmative but said that no reply had been received, although Sviridov had held out the prospect of a 25 percent reduction of the obligation.

With this interview the western trip came to an end. At home members of the delegation tried to put the best possible interpretation on it. Rákosi was probably the most upbeat among those reporting. It was the first time in modern Hungarian history, he said, that leading personages had, within two months, visited the capitals of four major powers. Among the losing nations Hungary was the only one able to do this and to give its hosts a firsthand account of its position. He added though that the terrible crimes committed by the Horthy regime could not be atoned for by a single informational trip. "The facts which have determined the Paris decision of the four foreign ministers cannot be altered by the most persuasive efforts . . ."[29]

Nagy hastened to write to Stalin offering an explanation why the trip had been necessary. As a rule such letters were forwarded through

Molotov or Dekanozov, but as both had gone to Paris, it was placed in the hands of a lesser official who promised to send it on. Nagy wrote that he had taken the delegation to Washington largely to seek support for the stabilization of the currency. He assured Stalin that his delegation did not, and would not in future, raise political questions in the west; in all such questions his government would adhere to the soviet position. "It knows that that is where it will find aid and protection."[30]

There are several evidences that Stalin was less than happy with this much-touted western trip, and especially with Rákosi's participation in it. Gyula Szekfü, who returned to Moscow after a three-week absence, reported that in his view "our position has weakened." A soviet deputy foreign minister who in previous talks had promised to investigate Hungarian grievances against Rumania now shrugged off another complaint with the remark that such excesses were due to the war and its aftermath. The Yugoslav ambassador, who reputedly got his information from very high quarters, indicated that the soviet government disapproved of the western trip and also looked with disfavor on the fact that Hungarian capitalists were allowed to profit from the inflation. At the same time Italian diplomats strongly hinted that at the peace conference the soviets would not support the Hungarian position against Czechoslovakia and Rumania.[31]

(Years later, when a Hungarian government delegation visited Moscow shortly after Stalin's death, Molotov coldly remarked that during the western trip Rákosi had been seeking a separate agreement with the United States. He hinted that Rákosi's inclusion in the delegation had greatly angered Stalin.[32] That the dictator nevertheless had not intervened is significant.)

When Nagy reported to parliament he admitted that he had not achieved the political goals he had sought in the west. He had not received promises regarding territorial and minority questions, but he added that "no representative of a great power will willingly undertake to advocate any change in decisions already taken."[33]

The aftermath of the western visit was as puzzling as its antecedents. On July 7 Sviridov presented Nagy with a series of demands aimed at weakening reactionary organizations, which continued to be a thorn in the side of the soviets. The government was requested to dissolve, among others, the Catholic Youth Organization, and to take strong measures against the Catholic clergy, as well as to expel and arrest a number of KGP parliamentary deputies.

Nagy bowed to some of these demands and refused others, notably the expulsion of duly elected representatives of the people. When the British and American envoys protested against Sviridov's high-handed conduct, he retreated and did not press the matter further.

The evidence clearly indicates that the soviet government still did not have a settled policy toward Hungary, or, alternately, that while it was ready to allow the government wide latitude in policy-making, it was constantly being challenged by antisoviet elements and organizations, as well as at times by the government itself. Stalin, so firm in his policies toward other satellite states, seemed unable to make up his mind how to treat the hard-nosed Hungarians.

Chapter 13

Preparing for the Peace Conference

Despite cumulative disappointments in the councils of the great powers, the general feeling in the country was that the peace delegation should argue its case vigorously and virtuously at the soon-to-open Paris Conference. The nation should not stand before history meekly accepting for the second time a crushing peace treaty the great powers imposed on it. Even if the dossiers of briefs and documents feverishly being prepared in the foreign ministry failed to impress the 21 nations whose delegations would assemble in Paris, they might be put to good use later when, in the incalculable course of European politics, fresh opportunities arose. Also, while the foreign ministers had not recommended border changes, the peace conference could, at least technically, still decide otherwise. Czechoslovakia in particular would appear before the conference in a less-than-favorable light, because of its flagrant conduct in minority affairs and its extravagant demands. The inequitable nature of the population-exchange agreement with Hungary was in itself proof of the republic's imperious ways.

The agreement had been signed on February 27 but was not ratified in Budapest until May 14, and then only by a narrow quorum and by voice vote. The debate had bristled with interparty rivalries. The Smallholders advocated rejection (although they knew that this would give the Slovaks an excuse to intensify their persecutions); they blamed the left bloc for the unhappy conditions in Slovakia.[1] Gyöngyösi urged his party to vote for ratification, arguing that it was the best agreement Hungary could get. He reasoned that the Czechoslovak claim that 300,000 Slovaks in Hungary wished to resettle could best be disproved if the agreement was allowed to go into effect and the number of Slovaks volunteering turned out to be far below Prague's estimate.

The controversy concealed a larger question that the great powers would have to decide: should they, after the bitter lessons of the interwar period, be guided by precepts that were most likely to lessen international tensions, or should they treat a country's claims entirely on the basis of its conduct before and during the war? It was obvious from casual remarks that the consciences of most statesmen pointed in the former direction, but they were invariably overridden by national interests. After the First World War there had been no "spheres"; Wilsonian "justice" was the only criterion for drawing borders and for protecting ethnics torn from their motherland. Yet it was precisely the misapplication of Wilson's lofty principles that led to World War II. Now, in the wake of that war, Hitler's former victims made sure that Germans would never again be a troublesome minority in foreign lands: they expelled them into a greatly reduced Germany. With the German question thus "solved," the peacemakers felt free to treat other ethnic minorities with benign neglect. Yet time was to show that the nationality question was not laid to rest through neglect.

Stalin sought to neutralize it by drawing East Europe into an ideological bloc. The western powers grumbled, but by the spring of 1946 they recognized the Moscow-controlled governments with some meaningless reservations. The Hungarian-Czechoslovak dispute remained unsettled however, and so did the position of each party in the developing global struggle. The multiparty governments in these two neighboring states and their free contacts with both the east and the west, made them candidates for a neutral bloc, unaligned, and open to capitalistic investment. Had they united in this stance, they might have been able to resist sovietization from the outside. But the ethnic issue divided them. This caused the western powers no end of discomfiture. They could not disinterest themselves as they had after Munich; at the peace conference they would have to favor one over the other and it was bound to be a difficult choice. Czechoslovakia was a Slavic state with a militant neo-Pan-Slavism as its national myth. It had an assertive Communist Party with a sizable plurality. In Hungary by contrast a centrist party had gained an electoral victory and public sentiment was unmistakably prowestern. On the other hand Czechoslovakia was a victor state and carried a halo as Hitler's long-suffering victim; Hungary had been a handmaiden of Nazi Germany and fought with it to the end. How to choose then?

A Hungarian legation officer in Washington sent the foreign ministry in Budapest an article from the pen of an American journalist based in Paris that analyzed Washington's dilemma. "American foreign policy," it argued, "seeking a spot in Europe where it can test its 'toughness' and strength, has looked on Hungary and decided . . . that 'this is the place.'" The author went on to explain that the United States, while proposing a three-power effort for Hungary's rehabilitation, had also launched a campaign on its behalf at the peace conference. The plan was to secure for Hungary territorial compensation for the Bratislava bridgehead, which was sure to be given to Czechoslovakia, and to defer indefinitely the question of the resettlement of Hungarians left in Slovakia after the population exchange.

The United States was aware of course that before the war Hungary was "the most feudalistic country in Europe and . . . during the war seized more territory than any other agressor [sic] except Germany itself." But Washington had to deal with current realities. As its ambassador in Moscow, Walter Bedell Smith, explained to a group of British and Dominion delegates: "The Communist Party is the third party in Hungary; in Czechoslovakia the first party; that is why we [Americans] favor Hungary." In American calculation soon after the conclusion of the peace treaty a large soviet army would be withdrawn from Hungary, which would greatly improve the prospects of the return on western capitalistic investment, prospects that were much better there than in Rumania or Bulgaria.[2]

The view from the Moscow legation was different. Gyula Szekfü had been in and out of the soviet capital and each time he returned to Moscow he perceived a change for the worse in official circles. In late June 1946 he encountered a definite coolness, all the more alarming because at the same time the visiting Czechoslovak ministers were given a demonstratively friendly reception. "Czech ministers were during the entire sports festival Stalin's guests in his grandstand. When Clementis arrived, Vishinsky walked with him across the airfield with his arm around him." Such small gestures in the soviet setting were magnified out of all proportion and were the subject of a great deal of speculation. "Impression is general," Szekfü continued, "that the USSR will not sacrifice Czechoslovak or Rumanian interests to please Hungary. Czech situation [is] particularly noteworthy. Czechoslovakia's position has declined since 1918–1919. They didn't dare to raise a word of protest against the annexation of Ruthenia, and although the Czechs living there were given the right of option, the exercise [of that right] is constantly hampered and even the transfer of their movable property is made problematic. They had to abandon the twenty-year old thesis of Czech and Slovak unity and settle for a dual government on the model of Austrian-Hungarian Monarchy and give special consideration to the Slovaks because any Slovak drive for independence would be damaging to Czech prestige."

Consequently Prague made the Slovak effort to expel the Hungarians its own. Moscow supported this position, although it was mindful of the fact that Czechoslovakia had, after World War I, been set up as part of the *cordon sanitaire* against the Soviet Union. All that now lay in the past and the republic had become a rampart against western influence.[3]

But no country could be a reliable rampart unless it had a socialistic government. The Czechoslovak people had taken the first step in that direction, of their own choice, when they gave their Communist Party an impressive plurality in free elections. Stalin nevertheless stopped short of all-out support for the mass expulsion of Hungarians. In this one question concerning East Europe he seemed willing to bow, in international councils, to the will of the majority.

If Hungarians indeed wanted to argue their case forcefully in Paris they needed spokesmen of outstanding stature and János Gyöngyösi was not one. He had little experience in foreign affairs, did not speak any major language, and had strikingly unoriginal concepts. That a man like him was allowed to hold the foreign portfolio for over two years showed either that the government did not truly believe that the person of the foreign minister made much of a difference (it made none in 1920 when men of wide international recognition argued for Hungary to no avail) or that it saw a certain virtue in appointing a humble man to represent it before the world, in contrast to the men of wealth and rank who had sat in Hungarian cabinets for decades before. The communists were perfectly willing to let a Smallholder put his hand to a treaty that they knew would by any measure be a crushing one. The Hungarian Communist Party (MKP), as well as the Social Democratic Party (SDP), and even the Peasant Party (PP), largely distanced itself from the peacemaking process. The Social Democrats had always been unhappy with Gyöngyösi but exerted no great effort to replace him.

On July 9 the parliamentary foreign policy committee heard Gyöngyösi's report on preparations for the peace conference. There was no objection to his mundane proposition that in drawing borders the unity of the ethnic groups should be the main consideration. The only debate developed over the composition of the peace delegation. Gyöngyösi proposed himself and two of his deputies, representatives from each of the coalition parties, and a number of experts. The envoys in Paris and London would be added as the need arose.

Voices were at once raised against sending a delegation of thirty to forty men when the country was in such a tight financial condition. Some questioned whether technical experts were more important than men with broad international connections and a knowledge of foreign languages. When the debate reached this point, Gyöngyösi emphatically urged the inclusion of a communist in the delegation. József Révai rose to the challenge with a sharp: "Are you doing us a favor? No, thank you." He read into the record the observation: "In the name of my party, party considerations do not come into play."[4] Gyöngyösi insisted that international considerations on the other hand demanded the presence of a communist at the peace conference.*

The preparatory subcommittee had already composed a draft of the Hungarian position. An eloquent introduction drew a connection between a satisfactory treaty and the future of democracy in Hungary. The great powers would be wise to establish fair borders, for otherwise "it is to be feared that democracy may appear to the Hungarian people as a system that cannot properly represent the interests of the nation; democracy will then be a victim of peace-making and its enemies will be the winners."

On border questions the draft was quite specific. It asked for 22,000 square kilometers of Transylvania to restore a rough ethnic balance. To the north, as Czechoslovakia was not willing to grant its Hungarian minority legal status, "she cannot insist on the territorial status quo."[5] In economic matters the draft asked for an international investigation of Hungary's condition. It also requested a suspension of reparations payments and a multiyear moratorium on the payment of capital and interest on all other international obligations. It called for the granting of foreign credit for the purchase of capital equipment. An inseparable part of the whole complex of needs was the return of goods taken to Germany and Austria. The part of the goods that had been used up should be compensated for.

A separate chapter dealt with Hungary's claims on Germany. It noted a credit balance of 1.05 billion gold marks in Hungary's favor from the wartime

---

*The Hungarian Minister in Paris, László Faragó, a Social Democrat, wrote to Árpád Szakasits on July 31 that it was impermissible that one party, the one yet that essentially dominated Hungarian politics, should be able to escape the political consequences of a tragic peace. "Please make it a party principle," he wrote, "that a communist should come, not as an 'expert' or a third-rate member, but someone with international recognition. If our 'sister party' is unwilling, please arrange for my recall. In such a case the whole delegation should be bureaucratic, without a political character."—Institute for Political History (IPH), Budapest, Hungary, Social Democratic Party (SDP) Files, Fond 283, 12/34.

payment schedule and 2 billion gold *pengös* representing the value of goods Hungary had contributed to Germany's military needs.[6]

None of these claims were unreasonable and none (except perhaps the claim on Germany) were impossible to fulfill. But such criteria were by now beside the point. The rapacity of the soviets in satellite countries had warned the western powers off placing any assets of great value into the hands of the governments that now ruled those countries. Nothing was safe from Moscow's greed and, as John Hickerson had pointed out to Gyöngyösi in Paris, as long as soviet demands on Hungary continued, the return of assets would make little difference.

Meanwhile relations with Czechoslovakia were going from bad to worse. Prague was probably encouraged by the reluctance of the western powers to give Hungary any support and seemed intent on producing some fait accompli. One action it took was broadly called re-Slovakization. Ethnic Hungarians were being offered a choice to either declare themselves Slovaks and retain their citizenship or to be evicted from their homes and deported. Some 400,000 Hungarians, weary of the harassment, accepted the first choice. This of course carried no guarantees for the future; people with Hungarian names might well be caught up in a new wave of xenophobia. Ernest Bevin, at a meeting of the foreign ministers on June 13, had already taken strong objection to the practice, calling it "deportation tactics," at variance with United Nations principles. He realized that the purpose was to be able to go before the peace conference claiming that there was no Hungarian question in Czechoslovakia and thus resist its introduction by the Hungarian delegation.[7]

Even Rumania drew certain conclusions from Hungary's vulnerable state and from the apparent failure of the western visit. Petru Groza and his colleagues assumed that the Hungarian delegation had gone to the western capitals, among other reasons, to complain about the maltreatment of its ethnic kin in Transylvania. The Rumanian press carried a number of obviously inspired articles of an accusatory tenor. They ruled out any compromise on the Transylvania question and assailed Nagy's government for not accepting without further debate the decision of the Council of Foreign Ministers.

The keynote to all this criticism was sounded by a prominent communist, Lucretiu Patrescanu, who in a speech at Cluj, long a center of Hungarian national and cultural life, threatened Hungarians in Rumania who failed in their citizenry duties, or who during the war had moved from Rumania to Hungary. Soon the entire press branded all those who pleaded for the revision of borders reactionary. This amounted to a condemnation of practically every Hungarian within or outside of Rumania. Several newspapers charged that Hungarian appetites were insatiable, that they reached beyond Transylvania to Moldavia for expansion. One paper went so far as to say that Hungarians mourned the collapse of Hitlerism because it meant a loss of support for Hungarian imperialism. Groza alone kept a cool head. Even he insisted, however, that inasmuch as the rights of the Hungarian minority were protected by law, any agitation for revision was uncalled for.[8]

The Czechoslovak government delegation that had gone to Moscow and was so royally treated even received license to resettle its "surplus" Hungarians; Dekanozov informed Szekfü of the decision. The fact that Foreign Minister Jan Masaryk, on a visit to London, failed to move Bevin from his opposition to deportations seemed of little consequence. *Moscow locuta, causa finita*—or so it appeared.

The possibility of a new war was all the time a live topic among diplomats and journalists alike. Two global conflagrations within a quarter century had accustomed the public to violent solutions to international problems. In the United States, by one report, features of a new war were being studied and concepts such as that the "northern corner" (the Arctic region) would be the strategic center and missiles would be capable of traveling 3,000 to 6,000 miles were taken quite literally.[9] Pronouncements by world leaders were being carefully studied and analyzed, and even when they were reassuring, the public remained apprehensive.

The ideological chasm was dangerous enough in itself; power realities made for a doubly dangerous world. The Soviet Union, backward and devastated, could depend only on its military forces and there was no telling when a crisis would arise that demanded it use that option. Former Foreign Minister Maxim Litvinov, in speaking to Walter Bedell Smith, said that "the best that can be hoped for is a prolonged armed peace." A month later he confided to a CBS correspondent he had known from the time he was an ambassador in Washington during the war that the hope that the communist and the capitalist worlds could live peacefully side by side was no longer realistic. "There has been a return in Russia to the outmoded concept of security in terms of territory—the more you've got the safer you are." "Suppose," the correspondent asked Litvinov, "the West were suddenly to give in and grant all of Moscow's demands. . . . Would that lead to good will and easing of the present tensions?" Litvinov's answer was that the result would be "the West being faced, after a more or less short time, with the next series of demands."[10]

Moscow was becoming a listening post of the first importance because, while it could not match Washington's broader representation and greater sophistication, its obsessive secrecy and impenetrable bureaucracy fostered the belief that the question of peace or war would ultimately be decided there. There was also the fact that many western statesmen and diplomats finally were getting a firsthand look at a society that had astounded the world with its heroic struggle in the Great Patriotic War. For Hungarians this was a particularly exciting opportunity as Russia had been a closed book to them even during the few brief years that they had had diplomatic relations with it.

In June 1946 a humble first secretary at the legation sent a lengthy and superficially exhaustive report to the foreign ministry in Budapest. Conditions in the USSR, he wrote, obviously called for peace. He followed up this statement with a nonsequitur. "Her internal condition is favorable as compared with the two Anglo-Saxon powers and she could undertake another major war. As we all know, England and the United State face major parliamentary struggles if they want to modify their country's social and economic structure. There is no such 'ferment' in the USSR." Soviet society was apolitical, the writer went on, a fact that was not due to its dictatorial system. There was only one party, not because there was a ban forming another, but because there was no need for one and none was desired. World War I and the Bolshevik Revolution had destroyed the old "intelligentsia" and the soviet government had at once addressed itself to creating a new one. The new intelligentsia had achieved full maturity when war broke out and the defeat of its major enemy was its first great achievement. It was apolitical only in the American and European sense. Its interests were focused not on questions that fell within the purview of the government but on those such as workers' competitions, the position of the trade unions, and social issues.

The government, the writer explained, was homogeneous. "The purposeful, unbroken, straightforward activity which Stalin had followed since Lenin is a guarantee that there can be no individual differences within the government. The rise of a [dissident] faction as had made the purges necessary is out of the question."

While the government did have the backing to lead the nation into war, social and economic conditions argued against it. According to the author of the report, the last war had arrested social development, especially the maturation of an intelligentsia that every country needs for its governance. The present leaders had achieved this maturation, but no government can in the long run be in the hands of "providential men," and the soviet state still had to develop a managerial class of full worth. "To use a technical term, it is not a finished product, only a raw one. Thus the USSR does not have full-fledged governance. It must develop one before the Old Guard dies out. For this it needs a long period of peace."

In foreign policy, the legation secretary continued, the goals were self-defense and the support of people's democracies. Anti-soviet circles in Europe and America saw behind every participation of the soviets in international affairs some secret machination advertising progress but seeking to achieve imperialistic goals. The matter was not so simple. Many motives fed into soviet politics: distrust of the western powers, memories of intervention in the Civil War, the *cordon sanitaire*, the atomic bomb, current antisoviet declarations. There was also the millennial impulse for expansion and the zeal of a newly emerged great power that wanted a leading role in international affairs. The new Russia saw itself as a liberator of peoples. (So, the writer might have observed, had Holy Russia of the past.)

Russian foreign policy suffered from many weaknesses. Its diplomatic personnel was half-trained and because of its long isolation lacked the smoothness and the air of intimacy that characterized European diplomacy. Tasks that other countries entrusted to diplomats, the soviets often assigned to the army; it was a matter of numbers, as there were a few thousand diplomats and millions of soldiers.[11]

What the writer refrained from saying was that it was precisely the reliance on raw force that made soviet diplomacy at once dangerous and ineffective. Diplomacy assumed give-and-take and Stalin's temperament inclined him only to take. If he compromised, it was only because the alternative to his position, which in these tense, Cold War years and the possibility of a new war, was unacceptable. But where questions of East Europe were concerned, he ruled out compromise from the start. In all fairness, the wartime spheres-of-influence arrangement Stalin had worked out with Churchill cut both ways. Stalin asked for no influence or privileges in West Europe; even in Greece, where a civil war continued to rage, he kept aid to the communist forces to a minimum. Conversely, he wanted no meddling in his sphere. If countries to which he granted a measure of freedom of action (Hungary in particular) took this to mean that they could cut themselves loose from the soviet bloc altogether, they did so at their own risk.

Chapter 14

The Ordeal in Paris

The two distinct voices that Hungarian diplomacy adopted in its balancing act between east and west found expression in the notes János Gyöngyösi addressed to the representatives of the great powers in anticipation of the peace conference. The only common plea was that a commission be formed to investigate and recommend a solution for "the Hungarian question." The notes sent to Arthur Schoenfeld and Alvary Gascoigne went on to say: "In presenting this request, Hungary looks with particular trust to those great powers which have . . . held a conference on the Crimean peninsula and have undertaken to establish a liberated Europe. . . . In the view of the Hungarian government this can be brought about in the most satisfactory manner if just and genuinely democratic principles, permeated by the Atlantic Charter and the San Francisco basic documents will find employment in international settlements." The note to Georgy Pushkin made no such embarrassing reference. It was humble and self-effacing. "Before the peace conference the much suffering and long misguided Hungarian people looks . . . with particular trust and hope to the Soviet Union which had never recognized the post–World War I settlements and whose wise statesmen, especially the great Lenin, have condemned the methods by which these peace settlements were constructed. Apart from these considerations, the Hungarian government asks for the special support of the Soviet Union for this reason too: it is the great power particularly interested in the Danubian basin and has a primary interest in [bringing about] a just settlement in it as well as frank understanding and cooperation among Danubian States."[1]

The argumentation in this letter might have carried some weight a year earlier but had lost its moral force since. Times of war are better suited for the promulgation of noble principles than periods of peacemaking, when hard national interests clash. References to the Atlantic Charter were out of date and the Declaration on Liberated Europe had been reduced to meaninglessness by soviet actions in Poland, Rumania, and elsewhere. As to Leninist principles, they had been overtaken by a new phase in soviet imperialism. The flux in East Europe called for an opportunistic diplomacy of which Stalin was a past master. He had installed in Bucharest a Moscow-friendly government (Petru Groza was a member of the leftist Ploughmen's Front, which was similar to the Hungarian National Peasant Party) and rewarded its subservience by protecting Rumania from Hungarian claims on Transylvania. Prague was induced to cooperate by soviet support for its Slavic aspirations. Even the Hungarians might reconcile themselves to the bitterly resented territorial status quo if they were in exchange allowed a measure of internal independence. With the Danubian states in line, the entire navigable portion of the river would be under effective soviet control.

But Hungary was once again an exception to this tidy scheme. Its deference to Moscow was entirely verbal. Nearly every speech by a public figure stressed the need for close cooperation with the Soviet Union, but the phrases rang false when the premier was a devout Catholic; the ambassador in Moscow (Gyula Szekfü) was a man who in his long career had served Habsburgs, Horthy, the Nazis, and now, like a latter-day Talleyrand, the Hungarian Republic; and the largest party in the country unmistakably spoke with the voice of reaction.

Hungary was a special case also because while victor states like Poland and Czechoslovakia and meritorious defeated states like Finland and Rumania had all lost land to the Soviet Union, Moscow raised no claim of this nature against Hungary. A quarter century of righteous indignation over Trianon served as a shield against further amputations. Even the Czechoslovaks, in the best position to demand territorial cession, asked for no more than a small bridgehead embracing five villages. They proved all the more stubborn and vindictive though in putting their Slavic national idea into practice. They evinced not the slightest interest in where the Magyars expelled from Slovakia would find a home; the Budapest government, as we have noted, was at a loss. Most of the confiscated Swabian farms had been given to tenant farmers who had not benefited from the land reform. For the Slovak side the fact that most of the Hungarians slated for

deportation were farmers promised to be a boon, because these unfortunates would leave cultivable land behind. This argument was of course never
used; only the ethnic one was. It almost seemed as though the Czechoslovaks had taken a leaf from Hitler's book and given their national policy a
racial character. The western powers pontificated and Moscow sided with Prague.
Vladimir Clementis's June visit to Moscow was followed by one from Communist Party chief Klement Gottwald in July. A communiqué issued at the
end of the talks stated that the soviet government was in general agreement
with the Czechoslovak position.[2] Gyöngyösi had a report from Washington
that the proposal for the forcible resettlement would be raised by Molotov
at the peace conference—an ominous prospect. Britain and the United States
were not likely to do more than protest because, according to the same report, Washington at any rate had given up on the strategy of "outwaiting"
Russia and wanted a speedy settlement on the dubious theory that "peace
begets peace," a phrase that, the author of the report noted with heavy sar-
casm, "can also be expressed as 'health cures.'"[3]

The peace conference opened in Paris on June 24; the first month was taken
up with procedural questions. This gave the foreign ministry in Budapest
more time to define its position. Gyöngyösi met with the parliamentary
foreign policy committee on August 3. He received instructions to reject,
with all due emphasis, the Czechoslovak charge that Hungary shared responsibility for the Munich decisions. That count of indictment could not
be allowed to form the basis of an adverse ruling in Hungary's dispute with
the republic. At a subsequent meeting, on August 5, Gyöngyösi, ever the
realist, asked for a drastic reduction of the claim originally advanced to part
of Transylvania. He had been to Paris to address the preparatory peace conference and found that his demand for 22,000 square kilometers had a frosty
reception. (He elicited more sympathy though when he pleaded that the
border should be so drawn that a roughly equal number of minorities would
remain on either side.)*

---

*In addressing the conference Gyöngyösi had presented some chilling statistics to
illustrate the impoverishment of his country. Before the war Hungary's national wealth
had been 52 billion *pengös* ($10 billion). Thirty-five percent of that had been de-
stroyed in the war. Agricultural loss was of the same percentage, but there was a 50
percent reduction of livestock. One third of the industrial plant was destroyed and
two-thirds of the rolling stock had been taken abroad. Present real wages were but
one-fourth of what they had been in peacetime. Reparations deliveries amounted to
40 percent of the GNP. As to manpower losses, of every 1,000 Hungarians, 23.5 were
war casualties.—New Hungarian Central Archives (NHCA), Foreign Ministry Files,
US Administrative, Box 171, 9/e.

Gyöngyösi now proposed that the claim be reduced to 5,000 square kilometers—otherwise Hungary could not depend on the support of a single power. The committee did not take a kindly view of this. Even members of the left bloc felt that the foreign minister was retreating too quickly. The Social Democratic Party (SDP) speaker wanted advance assurance that the smaller claim would be granted. From the communist side József Révai, usually critical of a revisionist program, this time wanted to know how Gyöngyösi justified reducing the original claim by three-fourths. All this of course amounted to hardly more than a holding maneuver because in truth everybody on the committee would gladly have settled for 5,000 square kilometers. In the end Gyöngyösi was given a free hand.

There was unanimity though in rejecting the Czechoslovak demand for enlarging the Bratislava bridgehead. In this position Hungary had the full support of the population of the five villages, which were firmly opposed to being transferred to Czechoslovakia.

Finally the committee discussed the selection of the delegation that would go to Paris. Gyöngyösi would head a party of 22 men. Privately a number of people on the committee and outside it had their doubts about Gyöngyösi's suitability for the role. He would have to match wits and prestige with men like Jan Masaryk and Georghe Tatarescu who enjoyed international renown. But there was no getting around the selection of the foreign minister to lead the delegation.[4]

On the domestic scene the left and the right continued to use foreign policy questions for political gain. The Smallholders posed as the sole advocates of border revision and cleverly implied that it could be brought about only with western help. The Hungarian Communist Party (MKP) asserted that American efforts to reduce Hungary's reparations obligations were intended only to ensure U.S. citizens heftier compensation for wartime losses. (When stripped to its core, this argument defended exorbitant soviet demands and confiscations.) The SDP tried to use its international connections to improve Hungary's chances at the conference. It cited its long and respectable history and its struggles, not only for socialism, but also for equal rights for minorities. Árpád Szakasits even made a trip to Paris to revive his party's contacts in the interest of Hungary's claims.

The Hungarian delegation arrived in Paris on August 11 and Gyöngyösi first addressed the plenary conference on August 14. The Czechoslovak and Rumanian chief delegates had already spoken. Jan Masaryk had made an emotional speech, reminding the delegates that they were called upon to draft treaties for defeated nations "whose Governments had, some for longer some for shorter time, but all for long enough, affectionately held Hitler's

hand during the unspeakable period when Nazis became the government of the so-called Third Reich and the second World War in our
generation was unleashed." His country's experience made it very
difficult for it to ensure the minority rights it had so generously
granted between 1918 and 1938. "What happened to Bohemia as a
result of her western neighbor and her German minority running
amuck, you know full well. What happened to Slovakia is also a
well-established historic fact. We in Czechoslovakia . . . have had
much more than our portion of scheming minorities."[5]

(It should be noted that Masaryk did not fully share the extremist
views of his colleagues and didn't even entirely subscribe to the views
expressed in his speech. Two days after the address he told a State Department official in the greatest confidence that he was willing to consider a frontier adjustment in Hungary's favor if that would enable his
government to resettle the ethnic Magyar bloc from Southern Slovakia.
The American was well disposed toward the offer and wrote home that it
"provides in our opinion the best possible solution provided the cession of
territory is adequate for this purpose."[6])

Tatarescu had spoken for a defeated nation and the bulk of his speech
was a long tribute to Rumania's armed forces, which, in their fight against
Nazi Germany, had performed great feats of courage and earned the recognition of even "the glorious chief of the Red Army, Generalissimo Stalin."
He touched on his country's historic conflict with Hungary only in passing.
He declared that the foreign ministers' decision to restore the whole of
Transylvania to Rumania "puts an end forever to the prolonged and renewed
oppression of which the Rumanian people had been the victim." He reminded his audience that the struggle of the Rumanian forces after his
country's surrender had been conducted against Hungarian as well as German forces. He begged the powers not to demand reparations from Rumania. It was Rumania who deserved reparations, from its former enemies,
Hungary and Germany.

Gyöngyösi's speech was longer than that of either of his adversaries but
it added little to the arguments all too well known to the assembled delegates. He too called attention to his nation's fight against Germany, but he
invoked the recognition only of Marshal Voroshilov, not the glorious Stalin.
He pointed to the strides Hungarian democracy had made since its birth a
year and a half ago. So far as minorities were concerned, the fact that Hitler
had misused the rights granted to them should not be reason to deny those
rights now.[7] Speaking of Hungarians under foreign rule, he said dramati-
cally that "the land and the people who tilled it for centuries and implanted

their civilization therein, are indissolubly linked together." He again called for an international commission to collect the necessary data to resolve the Hungarian-Czechoslovak dispute.[8]

After Gyöngyösi and his delegation left the room, Masaryk rose to speak. Referring to "the somewhat surprising and unprecedented declaration just made by Hungary, an enemy state," he asked for permission to study it, so that he could offer a reply. His request was granted.

Masaryk availed himself of the privilege the following day. He assailed Gyöngyösi's arguments and cited Hungary's feudal past. His position was supported by the next speaker, the Byelorussian delegate. Then Secretary of State Byrnes followed, but his remarks referred to the Italian peace treaty, which was being held up by the soviets' intransigent attitude. Andrey Vishinsky, who was the next speaker, chided Byrnes for dragging the case of another country into the debate concerning the treaty with Hungary. He explained why his country's reparations demands were fair and reasonable, calling attention to the fact that only a small portion of them had actually been paid.[9]

As was the case at prior international conferences, the great powers made sure that the pleas and speeches by statesmen of lesser countries had the minimum disruptive effect on the fragile consensus they had worked out among themselves. In fact one of the decisions made early in the conference was that the conclusions of the plenary assembly were subject to the approval of the Council of Foreign Ministers. A curious by-product of this decision was that questions of passionate interest to the inhabitants of a region were often entrusted to statesmen from countries far removed from it, in terms of geography as well as familiarity. The main committee investigating Hungarian border questions consisted, apart from representatives of the border states and the great powers, of delegates from Australia, New Zealand, Canada, and South Africa. The subcommittee dealing with the Hungarian-Slovakian dispute was made up of Australian, New Zealand, Canadian, Ukrainian, and Czechoslovak delegates, hardly an assembly to fill Hungarian hearts with confidence. To make the point even more emphatically, the Czech member was appointed reporter.

Gyöngyösi was particularly upset because the dispute with Czechoslovakia was the only one that still allowed some room for maneuvering. On August 19 he had word that the subcommittee would hear Hungary's case the next day; he secured a meeting with Secretary of State Byrnes to protest against the very short notice. Byrnes assured him that the Hungarian delegation would later have ample opportunity to argue its case. He also heard Gyöngyösi's complaint about the delay in restoring the displaced goods from Germany and replied that it was due to a common agreement among the victor powers not to return assets from their occupation zones to a country that was not a member of the United Nations.[10]

On this day, August 19, the subcommittee heard the Czechoslovak argument for the extension of the Bratislava bridgehead. The Canadian delegate spoke in favor of ceding three of the requested five villages, but only on condition that Czechoslovakia compensated Hungary elsewhere. This the Czech delegate categorically refused. The next day it was Hungary's turn to present its case. An English-speaking member of the foreign ministry, István Kertész, spoke for the delegation. He argued that the grounds offered by the Czechoslovaks for enlarging the bridgehead were inadmissible as the city had always grown in a northerly direction. He reminded the subcommittee that already in 1919 the Czechoslovaks had made a similar request, citing strategic reasons, and had been rebuffed. What could not be altered however was that the dispute was between a victor and a defeated state; after sundry proposals and counterproposals the subcomittee approved the transfer of three Hungarian villages to augment the bridgehead.[11]

Gyöngyösi's delegation was next called upon to argue its case vis-à-vis Rumania. On the afternoon of August 31 Pál Auer, Hungary's envoy in Paris, and a man of great international prestige, presented his country's position. The territorial claim on Rumania had by now shrunk to 4,000 square kilometers but Auer also requested wide local autonomy for Hungarians within the ancient Szekler enclave in the southeastern corner of Transylvania.[12] The members of the subcommittee, and even the Rumanians, acted as if this question had not yet been decided. Tatarescu argued on September 2 that the plurality of Hungarians in the disputed region was not great enough to justify the shifting of the border, especially as the economic unity of the province would be disrupted if the towns marked on the Hungarian map were detached from Rumania. He turned away a previous Hungarian suggestion that the commission mandate talks between Hungary and Rumania on minority questions as unwarranted interference in Rumania's internal affairs. He insisted that there was absolutely no discrimination against minorities.[13]

More and more it appeared that the Hungarian delegation, with its 22 members, was a perfunctory presence, just as the delegation in 1920 had been. In an effort to salvage a modicum of national prestige, Premier Ferenc Nagy traveled to Paris. He secured an interview with Walter Bedell Smith and somewhat transparently explained that he was trying to preserve in Hungary "the western idea of democracy" in contrast to the neighbor states in which "the eastern form of democracy" held sway. He referred to a "feeling of helplessness" among the Hungarian people and strongly hinted that unless the country received western support for the mitigation of the peace terms, he could not remain premier. That, he said, would have serious political consequences and might lead to civil strife. What he got in response was a collection of blank phrases. The United States, Smith assured him, "had no intention of receding from its previously expressed policy of assisting the peoples of eastern Europe to reconstruct their countries on a sound economic and political basis." He asked, obviously to appear helpful, what specific terms Hungary desired in the peace treaty.

Nagy asked for a slice of Transylvania "for political and psychological reasons." He wanted the conference to refuse the resettlement of 200,000 Hungarians and also the request for the Bratislava bridgehead "primarily for strategic and prestige reasons." The only encouragement Smith could offer was that the bridgehead issue "might serve as a basis for some give and take."[14]

Perhaps to soothe Nagy's feelings, Smith arranged a meeting with the secretary of state for the afternoon of September 7. The results were not happier than they had been in previous high-level meetings. Nagy's reminder that Hungary had had free and open elections whereas Rumania (and Bulgaria) had not elicited the reply that Byrnes "sympathized with Hungary's position and hoped to hear of progress made to overcome economic difficulties and further developments toward the attainment of political freedom."[15]

Ernö Gerö had been included in the delegation for his excellent connections but his only contribution was to arrange a meeting between Gyöngyösi and Molotov that produced little beyond platonic expressions of goodwill. That none of Gyöngyösi's arguments made any impression on Molotov became evident when the commission dealing with Rumanian territorial questions met on September 5 and the Australian delegate ventured that as the frontier with Hungary "had been a sore spot in Europe for years . . . some additional Hungarian centers might be incorporated in Hungary." The soviet delegate opposed the proposal and was seconded by the Ukrainian; the last attempt to salvage a part of Transylvania was dashed.

The reaction at home to these disappointments was bitter. The parliamentary foreign policy committee met on September 12 and a Smallholder member wondered whether the Hungarian delegation should not be recalled from Paris.[16] He ventured that the proper psychological moment for so doing would be when the conference accepted the Czechoslovak demand for the unilateral resettlement of the 200,000 Magyars from Slovakia. Révai at once argued that any premature recall of the delegation would hurt Hungary's position. It would make sense only if the government decided not to sign the treaty and that too was unacceptable because it would mean that the armistice terms would remain in effect and Hungary would remain under military occupation. Besides, the debate over the Czechoslovak demand required the presence of the Hungarian delegation.[17]

Szabad Nép struck an upbeat, if mournful, note when it wrote: "The Transylvanian border remains the Trianon border. Painful as it is, the one and a half million Hungarians in Transylvania will remain Hungarian even [though they are] beyond the border. Their national identity and the safety of their development remains one of the chief concerns of Hungarian democracy."[18]

In Paris the wrangling over the still unsettled expulsion issue continued. On September 16 Clementis, Hungary's nemesis, tried to assure the full committee that his government proposed to deport only those Hungarians who were neither of Slovak ancestry nor willing to identify themselves as Slovaks. The Hungarian speaker, Aladár Szegedy-Maszák, challenged Clementis's allegations and warned that the proposed resettlement would create a dangerous precedent. It could not possibly be carried out in a humane manner and would have grave consequences for Hungarian democracy.

At the September 20 meeting of the committee Vishinsky and Smith squared off. Smith called the Czechoslovak plan "more than unpleasant." The soviet delegate declared it justified and expressed himself astonished that Hungary, unlike other states with minorities in neighboring countries, was not ready to welcome its minorities.[19] The Yugoslav delegate adopted Vishinsky's position, while the British supported Smith. A deadlock ensued and a subcommittee of five members was set up to resolve it.

The Czechoslovak delegates, sensing that the trend of opinion was flowing against them, sought to lobby the British delegation, which they assumed was more likely to support them than the American. Masaryk spoke to Gladwyn Jebb and restated his willingness to grant Hungary some land in exchange for its readiness to receive its ethnic kin but asked Jebb not to make that position public.[20] Clementis pleaded with another member of the British delegation to sound the Hungarians on the conditions under which they would agree to the Czechoslovak resettlement scheme. The Englishman refused the request with the somewhat lame argument that Britain was not a member of the competent subcommittee.

The western delegations, seeking a way out of the conundrum, fell back on the useless formula of advocating direct negotiations. The Czechs and Hungarians, perhaps to demonstrate the futility of such talks, did hold a meeting on September 29 under Canadian chairmanship. Pál Auer forcefully presented his government's position. Any unilateral transfer of population, he argued, would have to be accompanied by a cession of territory. The Czech delegate refused to even consider such a solution. He stated blandly that the bilateral talks could produce results only on the basis of the Czechoslovak proposition. (It appears that Masaryk was isolated in his readiness to trade territory for ethnic purity.)

The talks led nowhere and the subcommittee met again the next day. The Australian and New Zealand delegates showed lively interest in the population question. They asked for statistics, compared prewar and postwar figures, wanted to know how the deportation of Jews had affected the picture, and finally asked what territorial compensation Hungary had in mind. After lengthy consultation with the experts the Hungarian delegation came back to say that it was willing to accept one-third of those to be evicted without land but wanted border rectification in the areas most densely settled by Hungarians to accommodate the remaining two-thirds.[21]

Possibly the subcommittee was only going through the motions and never really considered a transfer of territory. But neither was it ready to approve the indiscriminate transfer of hundreds of thousands of civilians against their will. It reported to the full commission in this sense and when the latter rendered its judgment on October 7, it rejected the Czechoslovak plan of resettlement but left Hungary's Trianon borders in place, not only with Czechoslovakia but with Rumania and Yugoslavia as well. It called on the Hungarians to enter into talks with Czechoslovakia "to solve the problem of those inhabitants of Magyar ethnic origin residing in Czechoslovakia who will not be settled in Hungary within the scope of the treaty . . . on exchange of populations." If within six months no such agreement was reached, Czechoslovakia (though not Hungary) had the right to bring the matter before the Council of Foreign Ministers.[22] The Czechoslovak government in addition had to pledge that none of its laws would discriminate "between persons of Hungarian nationality on the ground of race, sex, language or religion, whether in reference to their persons, property, business or financial interest." (It should be noted that the delegates of the three soviet republics on the commission voted against the last article.)[23]

The economic clauses, as the Hungarians learned later, had occasioned sharp exchanges among the great powers, especially between the United States and the USSR. The American delegate argued vigorously for the reduction of Hungary's reparations bill to $200 million. He went so far as to warn that if this suggestion was not accepted, the United States would see what it could do on its own to promote Hungary's economic recovery.[24] While on the surface this sounded like a generous offer, it had serious political implications.

The choice was between leaving Hungary destitute under soviet tutelage or to give her liberal economic aid but insist on supervising its distribution and use. The soviet delegate, in his reply, refrained from putting matters into such blunt terms. He nevertheless stated the obvious, that there were political motives behind the American proposition. The United States did not want good relations between Hungary and the allies nor did it truly want the restoration of Hungary's economy. The United States was trying to be magnanimous at the expense of the USSR, Czechoslovakia, and Yugoslavia; if it really wanted to help Hungary, it only had to return the displaced goods from its occupation zone.[25] In the end the commission rejected the American motion by a vote of 7 to 5.

Much of this skirmishing was pointless. The USSR had already decided to reduce Hungary's reparations bill to $200 million, though this was an empty gesture because what the soviet forces had carted off under diverse headings amounted to many times more, by some estimates to as much as a total of $1 billion.[26]

On the military side Hungary was allowed to have a land army of 60,000 men and an air force of 5,000 men. It could not manufacture self-propelled or guided missiles, atomic weapons, or any war material beyond its narrowly defined needs. Nor could it manufacture airplanes on the German or Japanese models. As to the return of prisoners of war, it had to make separate agreements with each of the victor powers; the cost of the repatriation would have to be borne by it.[27]

The Budapest government was further enjoined to restore the rights and assets of citizens and corporations of United Nations member states, in the condition in which they were in September 1939. Goods that had been confiscated or otherwise alienated had to be replaced and damages were to be compensated for up to two-thirds of their value in Hungarian currency. The government also was obligated to indemnify "persons, organizations or communities which . . . were the object of racial, religious or other Fascist measures or persecution."

On October 12 the plenary conference accepted these terms. The final text was still subject to approval by the Council of Foreign Ministers, which was to meet in December. It would also decide pending questions relating to Danube navigation and certain economic matters.

When Gyöngyösi reported to the parliamentary foreign policy committee on November 8, he admitted that he had gone to Paris with high hopes, because Hungary had made a new beginning, because the Atlantic Charter protected minorities, and because he fully expected that this time vengeance would not be a factor. On the other hand, Hungary had appeared before the conference with two heavy liabilities, the nature of its interwar governments and its unremitting revisionism. The most compromising feature of the latter was that Hungary sought its goals not in an enlightened spirit but in the service of reactionary circles. What had surprised him was that its responsibility for the war was not cited as a factor and was mentioned only in connection with the Munich Agreement, and then only because Czechoslovakia brought it up. What did count for a great deal was the question of which Nazi satellite changed its allegiance to Germany and when, and in this respect Hungary fared poorly.

The territorial hopes entertained by so many were disappointed. It had to be remembered though that in Czechoslovakia and Yugoslavia public opinion and the press had voiced demands for additional Hungarian land and these never saw the light of day, largely because of soviet opposition. With regard to the Rumanian border, Gladwyn Jebb of Great Britain had confided to Gyöngyösi that the Rumanian territorial commission had seriously considered border rectification in Hungary's favor but decided against it for three reasons: (1) Transylvania under Rumanian rule had proved a viable economic unit; (2) any border alteration would cut through vital rail lines; (3) the Rumanian government had issued broad measures of minority protection, creating the hope that this problem was under control.

That the Hungarian delegation did good work was evidenced by the fact that the soviet delegation supported it on many questions, especially economic ones. For instance it turned away Rumanian claims for compensation of losses suffered during Hungary's occupation of Northern Transylvania and supported Hungary's unchallenged right to assets located in the territory of allied and associated powers, including Czechoslovakia and Poland. Of particular note was the markedly changed attitude of a Canadian delegate who at the start of the conference frankly told Gyöngyösi that he found the Czechs far more sympathetic than the Hungarians because he had fought in two wars at their side against Hungary; yet when the Czechoslovak claims were being considered, this same gentleman pointed to the correctness and moral worth of the Hungarian position.[28]

All this was meager solace but, considering its past, Hungary could have done worse. The vehemence of its anti-Trianon position had paid some dividends after all. By denouncing its injustice, even in the face of its shameful wartime record, and successfully projecting the national outrage over it, any further truncation of territory was ab ovo rejected. The economic punishment was harsh but, given the dynamism of the Hungarian people, its oppressive provisions could be overcome in time.

Chapter 15

Let the Peace Begin

One positive development in the aftermath of peacemaking was the markedly improved relations with Great Britain. His Majesty's Government (HMG) had not yet extended formal recognition to the Budapest government and diplomatic contacts were at best correct. But Ernest Bevin had been a consistent opponent of the Czechoslovak plan of deportations, had supported border rectification in Transylvania in Hungary's favor, and had probably had a good deal to do with the pro-Hungarian leanings of the Australian and New Zealand delegates on the subcommittee dealing with territorial matters. London had a new representative in Budapest, Alexander Knox Helm, and shortly after János Gyöngyösi returned from Paris, Helm paid him a courtesy visit. He congratulated the foreign minister on the outcome of the conference and assured him that Hungary could count on British goodwill, both politically and economically. His government was ready to send £200,000 sterling worth of tool machinery to Hungary as a gift and was also interested in the conclusion of a trade agreement. Gyöngyösi, perhaps annoyed by Helm's inappropriate congratulations, responded with asperity rather than appreciation. He wanted to know why HMG had not supported the Hungarian plea for the easing of the reparations burden. Helm explained that his government was bound by the text adopted by the Council of Foreign Ministers. Gyöngyösi then pointed out that the obligation to indemnify United Nations citizens and enterprises was heavier than the reparations bill. Helm reminded him that there was an essential difference, namely that the assets owed to United Nations members would remain in Hungary and serve its economic life whereas the reparations transfers would all go abroad. The talk then turned to the recent currency stabilization, which, it appeared, had checked the spiraling inflation in Hungary. Gyöngyösi professed himself proud of that achievement but added that it had greatly lowered living standards, especially for civil servants. "Stabilization of currency for now also means stabilization of misery," he said.[1]

Helm refrained from touching on the ever-sensitive dispute between Hungary and Czechoslovakia, though it was of vital interest to the Hungarian foreign ministry, which could not know how long British goodwill for Hungary would last or how far it would go. A British diplomat stationed in Prague had visited Budapest and had an unsettling conversation with a ministry official. He had been sent to study the Slovak situation and he appeared conspicuously unsympathetic to the Hungarian position. He frankly admitted, when evidence to that effect was shown him, that the British embassy in Prague abetted the government there to deport the Hungarians. It was in line with British peace policy, the official averred. There would not be an agreement between the two countries as long as the minority problem persisted and the only way to remove it was to rid the country of ethnic Hungarians. He acknowledged that Trianon had not been a just treaty but he didn't think that this was a good time to revise it. The Slovaks would not hear of granting land to Hungary, nor of assuring minority rights in cultural matters. Thus resettlement was the only solution.[2]

In a formal sense the peace conference was still in session, and would be until the foreign ministers had adopted its resolutions. Their council met from November 3 to November 12. The peace conference already had reduced Hungary's obligation to indemnify foreign citizens from 100 to 75 percent of their actual loss; the council further reduced it to 66 percent, a figure still beyond the realm of feasibility. The vexatious Danubian question—who could navigate it, and under what conditions—was left to the consideration of the great powers and the riparian states. On November 24 Gyöngyösi, in a last gesture of protest, sent a telegram to the peace conference. "If peace treaty [is] to be finalized as per present text [as] accepted by the peace conference we must reckon with total collapse of Hungarian currency and economic life. Victor states, instead of considering position of each defeated state separately, have drawn up a kind of treaty which, while bearable for some defeated states, means unreasonable burdens for Hungary."[3]

It was too late. The great powers were not disposed to alter conditions they had arrived at after so much haggling and they had ample experience with similar claims of incapacity by stricken states that had proved exaggerated in the long run. In any case, Hungary's fate would depend not on the kind of peace it got but on the nature of its relations with the Soviet Union and, internally, on whether it could maintain an equilibrium between its contending political forces and prevent the domination of the Communist Party and the suppression of all opposition. That in fact became the principal issue once the question of the peace treaty was off the agenda.

There was a sense of urgency in the Hungarian Communist Party (MKP) to firm up the party's political control; the bulk of the soviet forces would soon be withdrawn and the political weight of the communists would proportionately decrease. Premier Ferenc Nagy, in anticipation of just such a development, held a confidential canvassing of public opinion (Arthur Schoenfeld in his report called it a "Gallup Poll") and concluded that if municipal elections were held at that time, the Smallholders' Party (KGP) would win a great majority, while the MKP vote would be negligible. Knowing that such an outcome would be unacceptable to the soviets, he decided to postpone the elections, much to Schoenfeld's displeasure. The latter complained that the KGP was not providing "the moral leadership necessary to prevent further encroachment by leftists."[4]

Seasoned diplomat though he was, Schoenfeld could not rise above his American preconceptions and get the feel of a political climate that had developed over decades of class struggle and that was aimed not at consensus, but at the absolute victory of one side over the other. Therefore the nationalistic platform, which the Smallholders now occupied, decidedly had the greater appeal. Irredentism, the cornerstone of that platform, had been the only program on which Hungarians of every station could agree, but, given the results of the peace conference, it was now a dead issue and each party formulated its own agenda to the near exclusion of others. Schoenfeld had a long talk with Mátyás Rákosi on November 30. The latter was blunt about the "necessity of elimination [of] power reactionary elements in Hungary." He explained to Schoenfeld (and the latter reported it to his superiors in the terse shorthand of telegraphic communications) that "25 years of Fascist control had left Hungarian people incapable [of] understand[ing] democracy and they would have to be educated to principles [of] democracy by force if necessary." Rákosi even accused the Americans of aiding reactionary elements in Hungary. Schoenfeld objected, pointing out that "it was often difficult to determine meaning attributed by communists to term reaction."[5]

Rákosi himself displayed some confusion, perhaps because he was unclear about what the soviets were trying to achieve in Hungary. Evidences continued to be ambiguous. Nagy had on a prior occasion remarked to Schoenfeld that "Leftist victories in neighboring states would render Hungary's position vis-à-vis USSR impossible,"[6] but this was standard alarmist talk for which Nagy had not offered any proof. Despite escalating Cold War tensions that gave spheres of influence added importance, the soviets still made no overt attempt to steer, let alone dictate, Hungarian domestic and foreign policy. At the meetings of the MKP's political committee there were repeated calls for stronger action against the opposition, but the political base for such action simply did not exist. Gerö stressed the importance of placing more communists in the foreign ministry, yet the ministry remained, even after the peace conference, a Smallholders' stronghold. Control over the police was another contentious issue. The communist minister of the interior, László Rajk, had begun purging police ranks of persons with a reactionary past, but, as a result of the agreement reached between the Smallholders and the Communists in the spring of 1946, this process was reversed. Rákosi stressed the need to "infiltrate, undermine and attack the right wing of the Smallholders' Party in order to promote their fall," but the infiltration remained ineffective. At the December meeting of the politburo Rákosi even lambasted the left wing of the KGP for failing to pursue the fight against reaction and blamed it for the continuing economic difficulties. It was these "anti-populist elements," he charged, who prevented the success of the struggle against the speculators and the curbing of black markets, with the result that the living standard of the working class remained unsatisfactory.[7]

Moscow's seeming indifference was unsettling for the MKP because nothing threatened its future more than leaving the country's political future to a spontaneous alignment of forces. Szekfü in Moscow received no guidance and no advice, except in an indirect way. On December 1 he sent a report complaining about the lack of sympathy that Vladimir Dekanozov, the acting foreign minister in Molotov's and Vishinsky's absence, was displaying toward Hungary, especially when compared with his attitude toward other soviet-bloc states. He singled out the still-growing soviet support of Czechoslovakia, a development that could have dire consequences. He had an explanation for this support: "Czechoslovakia exhibits strongly leftist policies, best shown by the nationalization of middle-size and even small businesses and her recent differences with England and America. Then there is Slavic solidarity of which the Czechs make the most."[8]

A change of political direction in Hungary began at the turn of the year and gained strength during 1947 but it was due to internal developments, with the soviets playing only an ancillary role. It is convenient to date gradual sovietization from this time, but not as a phase of a preexisting program; there is no shred of evidence that the soviets had such a program. What came to be known as the Kovács-Nagy affair did draw in soviet police authorities but their intervention had no visible political consequences.

On New Year's Eve of 1946 the ministry of interior announced the discovery of a wide-ranging antigovernment, antirepublican conspiracy. Already some hundred people had been arrested in connection with the charge. As the ministry explained it, a group referred to as the Hungarian Unity Society had been recruited from diverse elements, but mainly from rightist Smallholders ranks, assembled a large cache of arms, and conspired not only against the republic (which it planned to overthrow to return the country to the regency) but also against members of the Red Army. It was this latter charge, with some foundation in fact, that brought the soviet military police onto the scene. In the center of the conspiracy stood the secretary-general of the KGP, Béla Kovács, the man who the past May had played a key role in forcing the ministry of interior to back down from its nearly exclusive control of the police. Kovács was however a member of the National Assembly and as such enjoyed parliamentary immunity. The MKP immediately brought pressure on the Smallholders to disown Kovács and lift his immunity. Nagy, and the KGP political apparatus, refused the request.

The State Department in Washington viewed these developments with unease. It saw in the KGP the best prospect of keeping Hungary out of the soviet grip, but now the very existence of the party seemed to be threatened. The head of the European Affairs Office wrote to Dean Acheson that "the Party seems badly shaken and confused and the Prime Minister has thus far evinced little intention to take a firm stand against these new encroachments." For the rest: "The Prime Minister has denied that soviet authorities have intervened in any way in the present crisis, but it is very probable that Soviet influence in support of the Communist drive is fully operative behind the scenes and that the Prime Minister has made this statement under some sort of personal pressure."[9]

The new secretary of state, George Marshall, cabled Schoenfeld an urgent message: "We are conscious [of the] importance [of] buttressing U.S. political support of democratic elements Hungary by material aid . . . and are earnestly endeavoring find ways and means to make such economic assistance possible without delay."[10] Schoenfeld offered his own grim assessment:

> It has been clear for a long time that if non-Marxists among Hungarian leaders do not find within themselves resources of character and political will . . . to oppose successfully encroachment of Communist monopoly it is also because they are obsessed by their identification of that minority with Soviet power which they consider irresistible.[11]

Juridically the Kovács case was still in its early stages but on January 23, in the MKP political committee, Rákosi was already contemplating the coming trial. He gave instructions that the audience should be carefully selected for political correctness. "Acceptable people's judges" had to be chosen for the bench. The minutes of a later meeting record him saying: "We must continue struggle to lift Kovács's parliamentary immunity. The left must be mobilized to declare that it holds Kovács guilty and so inform the President and the Premier."[12] Two days later the committee instructed Rákosi to make a radio speech and state that the interests of democracy demanded the liquidation of the conspiracy and "the turning back of every effort to cover for the conspirators."[13]

It was amid these portentous events that the foreign political committee of the National Assembly met to empower the cabinet to ratify the peace treaty. Gyöngyösi's report, although nonpolitical, raised the question by implication whether recent developments in Hungary had not emboldened the Czechs to pursue their program of forced resettlement despite the decisions of the peace conference, possibly relying on soviet support. He reported that the Prague government had issued a decree obligating citizens to perform public works; evidently the purpose was to uproot Hungarian families and send them to the Czech parts of the republic. The decree placed before many families the painful choice of either accepting transfer to Bohemia-Moravia or moving to Hungary voluntarily. That the purpose of the decree was not to gain manpower for public works was evident from the fact that not only able-bodied men but women and children were also subject to it. Gyöngyösi had sought a meeting with Vladimir Clementis but the latter had proved dilatory, leading Gyöngyösi to believe that the Czechs wanted to complete the displacements before engaging in substantive talks.[14]

In view of the gravity of the problems involved, the foreign policy committee decided to leave the decision regarding ratification to the full session of the National Assembly.[15] The government meanwhile suspended the resettlement of the Slovaks who had volunteered for it; thus the population exchange that was intended to be the first phase of turning Czechoslovakia into a purely Slavic state ground to a halt. Foreign envoys were beating a path to Gyöngyösi's door to be informed on up-to-date developments.[16]

On January 17 the British proposed to Washington that Britain and the United States, together with the Soviet Union and France, form a commission to supervise the population exchange. The State Department used legal arguments to parry the invitation, which, if accepted, would involve the United States in a part of Europe that was only of commercial, not political, interest. "Pending the coming into force of the peace treaty," the reply to London read, "and pending the outcome of direct discussions now apparently contemplated, it is the view of the United States Government that it is inadvisable to consider Four-Power action along the lines suggested by the British Government."[17]

At the same time the British ambassador in Prague, Philip Nicols, sent a reassuring report to London explaining that the internal resettlement was being carried out humanely and with the sole purpose of easing overpopulation in some areas and relieving labor shortage in others. This report apparently soothed the moral misgivings of the Foreign Office because when the Hungarian envoy in London, László Bede, went to see Bevin, the latter turned away his pleas for intervention by suggesting that the Hungarian government approach other powers; he probably had the Soviet Union in mind.[18]

It was by now a commonplace that American commercial interests in Hungary were closely linked to the survival of the Smallholders' Party as a potent political force. That in turn seemed to depend on the country's economic prospects, which at present still appeared gloomy. The State Department on February 1 worked out a three-phase plan for improving the outlook. It doubled the credit limit granted to Hungary from the present $15 million to $30 million and offered additional credit for cotton purchases, which, a note to the Budapest government stated, could "probably be arranged within a few weeks." It also sent a request to Congress to consider a relief grant to Hungary after UNRRA aid had expired "as a matter of urgency."[19]

But events were moving too fast for such sluggish measures to have a tangible political effect. The conspiracy case was being pursued with all possible speed. Kovács volunteered to appear before the political police for interrogation but would not relinquish his immunity. The soviet military police was by now several steps ahead of this position. On February 25 it arrested Kovács and subsequently took him out of the country. This was the first open intervention on the part of the soviets in Hungarian internal politics and a violation of parliamentary immunity. Georgy Pushkin, in an attempt to forestall a government crisis, told Gyöngyösi that he saw no reason for the cabinet to resign because of the arrest of the secretary-general of the chief coalition party. He did take the occasion though to complain about the "cool attitude" of the Smallholders toward the USSR and he blamed, as usual, reactionary elements.[20]

Secretary of State Marshall, although his own department conceded that the conspiracy charges had concrete foundation,[21] took a serious view of the soviet action. He instructed General George Weems on the Allied Control Commission (ACC) to lodge a protest with its chairman, General Vladimir Sviridov.[22] In a follow-up telegram the secretary of state wrote:

> These developments, in the opinion of the United States Government, constitute an unjustified interference in Hungarian internal affairs the effect of which will be to support the efforts of small group in Hungary to substitute a minority dictatorship for responsible administration representative of the will of the Hungarian people.[23]

Sviridov's reply to the note confirmed that an "anti-constitutional and anti-republican" conspiracy in fact existed and that most of the conspirators came from Smallholder ranks, as officials of the party themselves admitted. The USSR had the right to protect its armed forces on foreign soil. On this basis Sviridov rejected Weems's protest.[24]

The Hungarian press published both Weems's note and Sviridov's reply. Schoenfeld urged Washington to keep public interest in the case alive. Sviridov, he wrote, left key portions of Weems's note unanswered, notably those stating that the investigation was being conducted entirely by the communist-controlled political police, that the communists had scotched Smallholder attempts at a parliamentary investigation, that the so-called people's courts were also communist-dominated, that confessions were extracted under duress, and that Kovács had been arrested by military authorities when attempts to lift his immunity failed.[25]

A second note from Weems drew a curt rebuff from Sviridov, who claimed that the note did not contain any questions that he, Sviridov, had not addressed previously. He concluded, "I do not find it possible to change my attitude to your offer of mutual investigation of existing situation."[26]

As some feared, British and American interventions did more harm than good because they confirmed the MKP perception that the KGP was a foothold of western influence. At a meeting of the politburo Imre Nagy, an illustrious member of the Moscow-bred cadre, offered an impassioned foreign policy report. After accusing the United States of seeking to squeeze soviet influence out of people's democracies, he labeled the Kovács conspiracy "an attempt to restore, with the active support of foreign reaction, the old unbridled power of [domestic] reaction." He charged the conspirators with trying to come to power through an armed uprising. "In this connection it becomes clear that one of the most important supports of Anglo-Saxon imperialism in Hungary is the Smallholders' Party. Our policy is to unmask this party as a nest of Horthyist reaction, to shake its influence on the masses, especially the peasants, and to replace a government under Smallholder leadership with one under democratic leadership."[27]

Premier Ferenc Nagy tried to soothe the growing tension by sending Rákosi a solemn "Declaration" adopted by the political committee of the KGP, stating that the party was seeking to reform itself to bring an end to the crisis. It was ready to return to the radical program of 1930 (a time when it did not bear the reactionary stigma). He pledged a return to a less compromised program aimed at uplifting the working people and placing members of that class into the top ranks of political and social organizations, while not ignoring the intellectuals, the small merchants, or the traders either. It would afford the peasantry enjoyment of every opportunity that a democratic system could offer.

The same note also carried the committee's deep condemnation of the conspiracy and demanded that everybody proven guilty should be punished. "Together with the parties of the working class [the Smallholders] undertake the struggle against the excesses of capital and against the impudent and anti-national upsurge of wholesalers."[28]

Perhaps on this basis some accommodation could have been achieved, especially as the MKP still did not have the unconditional support of Moscow and had to rely on its own propagandistic devices for popular support. But just at this time a new phase of the Cold War opened and it sharpened the struggle between left and right everywhere. A new session of the Council of Foreign Ministers met on Moscow on March 10. It was the first one at which the United States was represented by George Marshall. Stalin was ready to receive him but Marshall at first refrained from seeking an interview. Possibly he wanted to await the outcome of the conference or perhaps he wanted to gauge soviet reaction to what he knew would be an explosive new development in the Cold War.

On March 12, in a speech to Congress, President Truman announced American determination to assist any nation that was resisting internal or external subversion or any threat to its independence. He asked Congress to allocate $400 million in economic and military aid to Greece and Turkey, two states who claimed that they were under heavy soviet pressure. The broad context of the speech made it clear that it was not a limited aid package but a global declaration of war on communism.

Gyula Szekfü wrote from Moscow that the Kremlin saw Marshall's hand behind the presidential declaration and that it was delaying progress at the foreign ministers' meeting until public reaction to Truman's speech could be ascertained. The soviet expectation, according to Szekfü, was that the United States would soon be engulfed in a deep economic depression and the public would then resent the financial obligations the Truman Doctrine involved.[29]

In Hungary, the political right gained new confidence from the speech. Its perception was that the government was being assailed both by internal and external subversion and PresidentTruman had pledged support against such attempts. When the perennial American presidential candidate Harold Stassen visited Budapest, Nagy told him that the growing decisiveness of American foreign policy had awakened hopes in every oppressed nation in East Europe. He conceded though that Smallholder politicians had concluded that it was unwise to accept the communist challenge and for now refrained from raising their voices.[30]

It was now the right against the left; the center had become irrelevant. This was even true for the embattled Social Democratic Party (SDP), which was being pushed into a centrist position. The British and the Americans had cautious hopes that the right wing of this party, with a program similar to that of the British Labour Party, would remain independent of the communists. But the SDP was just the party that most powerfully felt pressure from the extreme left and its choice was increasingly between merging or being branded as an opposition party.[31]

Rákosi had always had a grudge against the SDP; it was, as he saw it, halfhearted in its commitment to the workers' cause, and to the larger struggle against the reaction, which it had in fact served for three decades. The international prestige it enjoyed was in contrast to the mistrust communists encountered everywhere. He confessed his true feelings toward the "sister party" in a speech to a communist cadre in which he discussed the history of the MKP.

> Twenty-five years have deeply imprinted themselves on our comrades. No man can for 25 years be underground without consequences: he gradually becomes unaccustomed to working with the masses, he gets used to conspiratorial work, to propagandizing instead of persuading. ... The Social Democrats to this very day have an advantage in [the practice] of working with the masses, in labor unions and similar places, where they are in their element, one in which they have dwelt for 25 years unmolested. They have many advantages over us and we have to make great efforts to overcome this handicap, to shake off every remnant of 25 years of illegality, every obstructing feature of it.[32]

Rákosi might have been talking about himself; an indefatigable speechmaker, he remained a propagandist and his methods were eerily reminiscent of his underground past. More than anyone else he was responsible for the overheated political climate and the dictum that everybody who was not with the communists was against them. The SDP did not know where to place itself in the growing conflict between east and west—loath to sever its western connections, it could not pose as a working-class party as long as it maintained them. Since liberation it had claimed to be a link between the center and the left; free of the unsavory reputation of the communists, it relied on respectability and its professed commitment to internationalism. But such policy in the Cold War was a paradox. In crisis conditions European politics always tended to the extremes. The conspiracy case, which the communists fully exploited, created just such conditions. The republic was in danger and every party had to stand up and be counted. This was what the MKP demanded; to shirk this duty meant flirting with the charge of treason.

Szekfü in Moscow was experiencing a new spell of pessimism. The promising development in Hungarian-soviet relations that had begun with the Moscow visit a year before had lost its momentum, he wrote. "We had every reason to believe after those talks that we had entered the road traveled by other 'satellite' states such as Finland, Rumania and Bulgaria and that our burdens will be eased." But then contacts with soviet officials became formal and strained. One possible reason was that Ferenc Nagy's delegation, after returning to Budapest, had made exaggerated statements, allowing Hungarians to believe that Moscow would support their claims on parts of Transylvania. Subsequently, after hardly a decent interval, the delegation visited western capitals; that too was viewed unfavorably in Moscow. The soviets judged it "swing politics."

The soviets had long suspected, Szekfü averred, that Hungary was hoping for "a miracle from the Anglo-Saxons." The middle class and upper classes continued to propagate old clichés and Cardinal Mindszenty was in the center of resistance to progress. The discovery of the conspiracy confirmed these suspicions. Thus, Szekfü wrote,

> From all of the above it is clear that the greatest determinant of our relations with the Soviet Union is our internal politics. Obviously after the lost war we can no longer hope for an entirely independent foreign policy. We must always keep soviet interests in mind. Whoever thinks otherwise fools himself. Even if the Red Army leaves Hungary, it will stand on the border and the Soviet Union has treaties with Czechoslovakia and Yugoslavia whose armies complement the Red Army. With the German and Austrian peace treaties still pending,

the soviets cannot let their guard down and let go of any state in their security zone between Finland and Bulgaria.[33]

Meanwhile the Hungarian peace treaty received the blessing of the Council of Foreign Ministers and on February 10, in Paris, Gyöngyösi signed it amid much ceremony. French Foreign Minister Georges Bidault, presiding, might have been guilty of unintended irony when he said: "In my capacity as chairman I express, in the name of the delegates here assembled, my good wishes to the Hungarian delegation on this memorable occasion." He added later: "The documents before you put an end to the saddest and most horrible war. It is my firm hope that they will open the road to international understanding and conciliation."[34]

The Hungarian delegation obviously had no such hope. Following the signing one member handed Bidault a note protesting against the treaty's failure to ensure, for Hungarians living outside the country, full civil rights.

The treaty had a bad reception at home. Representatives of various parties used the unlikely opportunity of a debate over the foreign ministry's budget to voice their disappointment. The speaker, a Smallholder, bitterly remarked that in times like these small nations were not allowed to shape their own future. A discussion over relations with Czechoslovakia soon turned to wider issues. Dezsö Sulyok, a deputy from the right wing of the KGP, who had recently broken away from the party in protest against its fecklessness in the face of the left bloc and formed his own, the Freedom Party, warned of the danger of reawakening nationalism and "self-tormenting masochism." Hungarian public opinion had to be educated in the principles of Realpolitik, he said. Unfortunately the provisions of the treaty did not conform to those principles. He made a motion that the National Assembly enact a law providing for Hungary's eternal neutrality.

While hardly a realistic proposal, it had the virtue of being a compromise between the Smallholder and the communist view of the direction in which Hungarian foreign policy should proceed. A prowestern policy being impractical and a prosoviet one unpopular, neutrality was the only stance with which every party, even the MKP, could make its peace.

But at this crucial point the MKP deputies shifted the debate away from the peace treaty, which wasn't on the official agenda, to the Kovács affair, or rather to the recent exchange of notes between Hungary and the western powers. One deputy proposed that the government reject the western notes as interference in Hungary's internal affairs. The SDP and the National Peasant Party (PP) supported the move. These left-bloc parties then composed a joint letter, which they sent to the premier. They called on him to issue a communiqué in reply to the latest American note. They said that such a note

should emphasize that the judicial proceedings against the conspirators con-
formed to republican legality and that the coalition parties were in agreement that the conspiracy had to be liquidated.[35]

Nagy refused to issue such a communiqué. At a cabinet meeting he
argued that Hungary should not compromise its relations with the United
States with such belligerent messages. The MKP viewed the refusal as proof
that the premier was pursuing his party's, not the nation's, foreign policy.
The MKP political committee, during its March 20 session, resolved that
Hungary should, immediately after the ratification of the peace treaty, adhere to the bloc composed of the USSR, Czechoslovakia, Poland, and Yugoslavia (of which Rumania and Bulgaria were also in the process of
becoming members), for otherwise "we will become isolated from the truly
democratic countries and Hungary would become a base for Anglo-American
imperialist circles."[36]

The coalition concept lay in tatters. Until now only the deeply felt need
to secure a respectable peace treaty had given it some tenuous unity. Now
foreign policy became a function of domestic policy in an extreme sense.
The fear of offending the United States translated into devotion to outdated
concepts, a prescription for social retardation—this at any rate was the MKP's
view. In KGP eyes an eastern orientation meant total surrender to soviet
domination. There was some idealism behind both positions: the right, as
recent events showed, was willing to jeopardize its very existence, and its
members their personal freedom, to preserve ideals bound up with Hungary's
heroic past; the left, or a good part of it, had given evidence over many
years of its willingness to suffer illegality and persecution for a different
future for Hungary. That the conflict had an ugly feature was also bound up
with Hungary's past: the bulk of the MKP's leading cadre was Jewish, as
were many of the recently joined functionaries, and a majority of the hated
political police. This alone confirmed in the public mind a lingering conviction, one which Hitler had so successfully exploited, that Jews were a people
without a country, ready to turn against the nation that had given them a
home, in the name of the fuzzy internationalism promoted by Karl Marx.
The party's hostility to the Catholic Church was seen as proof of its god-
lessness and a rejection of divinely ordained truths.

In March and April of 1947, as the conspiracy case was acquiring ever
new dimensions, a national debate arose over the future of public education
and the KGP seized the opportunity to make its position unmistakably clear.
Nagy lent his name to a circular that was sent to all KGP organizations and
became a clarion call for rightist defiance. It stated that "our party will not,
in exchange for any political gain, take a step which would curtail our

religious congregations, which had become one with Hungary's history, her sufferings and joy. And yes, we are in favor of free voluntary religious education."

The circular continued: "We know that our nation—not only the peasantry but the great masses of industrial workers as well—are firm in their belief in God and in their devotion to their churches. In regard to religious education they follow not only rules and laws but their own consciences as well. Be patient and confident."[37]

Nagy had by now made one more concession to the left: he had issued a mild rebuke to Washington for its intervention in the Kovács affair. But Aladár Szegedy-Maszák explained to the State Department that the note proved Nagy's inability to resist soviet pressure, an effective loss of his freedom of action. The American official Szegedy-Maszák spoke to, Walworth Barbour, replied that to act on such a statement would be premature and would complicate Nagy's efforts to restore his authority.[38]

Reporting to Budapest, Szegedy-Maszák speculated that the proclamation of the Truman Doctrine had been "organically preceded" by the American note to Budapest concerning the conspiracy case. Acheson himself, who had revised Truman's speech to Congress, acknowledged the connection. Speaking to the press at the State Department, he stressed the effect of Truman's speech on East Europe and mentioned that Hungary's freely elected government was in danger. "I believe the principle is clear," the envoy quoted him as saying. "The President of the United States announced that [he and the American people] were concerned when a nation enjoying free institutions is forced to abandon them. . . . I have stressed before and we at the embassy [in Washington] have already protested against the action that is being carried out in Hungary."[39]

An exhausted Ferenc Nagy left for a Swiss vacation on May 14. When Schoenfeld visited him the day before he found the premier reserved about Hungary's prospects as an independent nation. "Hungarian Government would more successfully represent natural wishes and desires of vast majority of people," once Russian troops were withdrawn, Schoenfeld reported him as saying. Nagy foresaw a new crisis over two issues: foreign policy and the nationalization of the great banks, which the communists were pressing for. He was confident though that the crisis would not come to a head during his absence.[40]

This hope was rudely disappointed. At Schoenfeld's urging, Nagy had, before his departure, written to Sviridov, asking that Béla Kovács be returned to Hungarian jurisdiction. Sviridov refused, stating: "Competent soviet authorities have not yet completed investigation." On the evening of

May 28 Rákosi, as acting premier in Nagy's absence, presided over a cabinet session. He revealed the contents of Sviridov's note and added that the Russian had placed at his disposal information obtained during the interrogation of Kovács and his coconspirators.[41] The information implicated Ferenc Nagy. The cabinet decided to call for Nagy's immediate return.[42]

The next evening Rákosi brought the matter before his politburo. It decided that if Nagy returned he would be called to account for his activities and that the KGP should be informed of that at once. The MKP should establish liaison with other parties in the left bloc for the purpose of a common policy.[43]

Schoenfeld reported on May 30 that according to a Smallholder politician "alarming rumors were in circulation," including one to the effect that Nagy would be obliged to resign because Kovács's testimony implicated him.[44] Schoenfeld went to see Rákosi for clarification. Rákosi politicized about how the forces of reaction were still "very tough" in Hungary and unwilling to surrender. Schoenfeld's report ended with a weighty paragraph: "Speculation in political circles is that culminating phase of Communist offensive will now begin for decision of question of control of this country's power to end occupation."[45] On May 31 Rákosi held a brief cabinet meeting to deal with the question of Nagy's resignation. Gyöngyösi was absent, allegedly because of ill health. Rákosi announced that he had been present in the premier's office when State Secretary István Balogh received over the telephone Nagy's resignation from Berne. The written resignation was expected to arrive that day. The cabinet must also resign but carry out its duties until the president of the republic had released each of its members from his obligations.[46]

Chapter 16

Changing of the Guard

When Ferenc Nagy learned of the cabinet's decision to recall him, he first signaled assent, mainly because his young son had remained in Budapest. János Gyöngyösi then telephoned him in secret and suggested that Nagy contact the Hungarian Minister in Berne, Ferenc Gordon, first. After talking to Gordon, Nagy realized the gravity of his position and decided not to return home after all. He set conditions for sending a written resignation though: he should not be dishonored before the nation, his wealth should be left in his name because confiscation would be equivalent to a loss of honor, and his son should be flown out to join him and his wife.

Mátyás Rákosi handled the matter at the other end from this point on. He telephoned Gordon and criticized Nagy's conditions, especially the fact that he tied a public matter, his resignation, to a private one, concerning his son. Rákosi ordered the Hungarian Telegraph Bureau (MTI) to issue a communiqué to the effect that Nagy had resigned the permiership in Berne. He then called Gordon and insisted that Nagy must write out his resignation, addressed to President Zoltán Tildy and to himself and send it *en clair*; he should also send a copy to the foreign ministry in code. The argument went back and forth over the wires but the outcome could not be in doubt. Tildy, utterly intimidated, seconded Rákosi's demand.[1] Nagy in the end gave in and became the first major political figure to go into exile since the war.

There followed a spate of resignations among diplomats abroad. Aladár Szegedy-Maszák appeared at the State Department on June 2 and delivered a note the gist of which was that the government in Budapest was not "a free agent any longer." He added orally that he had been ordered by that government to return to Budapest "for consultation," but that he didn't intend to do so. As he did not recognize the present government he would not tender his resignation to it but would simply leave the legation and informally turn it over to a counselor, Pál Marik. He inquired whether he could count on "the benevolence of the United States to permit him to remain in this country."[2] His request was granted.

Arthur Schoenfeld had been relieved of his post in Budapest shortly after the Nagy affair broke and Szegedy-Maszák expressed his hope that the United States would signal its condemnation of Nagy's ouster by refusing to appoint a replacement and possibly go so far as to recognize Ferenc Nagy as the head of a government in exile. He was distressed to learn during a visit to the State Department on June 6 that a successor, Selden Chapin, had been appointed minister in Budapest and would proceed to take up his position.[3]

In the wake of Nagy's resignation the cabinet in Hungary was reconstituted. Another Smallholder, Lajos Dinnyés, became premier. Gyöngyösi was replaced, on an interim basis, by another Smallholder, Ernö Mihalyffi, who simultaneously held the propaganda portfolio. Dinnyés presented his program to parliament on June 10. He pledged that his government would respect private property and protect individual freedoms and rights. Externally, it would foster friendship with neighbor states and strive for the closest possible relations with the great powers.[4]

The last point might have been an attempt to defuse the tension created by the Nagy affair, as protests from the west were coming in a continuous stream. There were repeated references to the Yalta agreement and how its principles had been violated. U. S. Secretary of State George Marshall directed the legation personnel in Budapest to make it clear in conversations with government officials that the fact that the United States continued relations with the government "in no sense implies approval of situation resulting from recent events in Hungary or methods used to bring present government to power."[5]

Marshall proposed to bring the entire affair before the Allied Control Commission (ACC) and solicited the advice of the British government and of the American chargé in Moscow. Both replied discouragingly. The Foreign Office feared that if the soviets produced documents proving Nagy's involvement in the conspiracy there would be no means to discredit it.[6]

Another, undoubtedly well-founded, fear of the Foreign Office was that soviet rejection of parallel notes from the British and American governments would only demonstrate the impotence of the western powers. George Weems was nevertheless instructed to deliver a protest note to Vladimir Sviridov; this he did on June 10. Sviridov in his reply termed the accusations contained in the note "unfounded fiction." He asserted that the crisis was caused by Nagy's refusal to return home; the new government had been formed "in strict conformity with constitutional standards." Moreover, "the previous distribution of portfolios among the coalition parties . . . remained basically the same."[7]

Sviridov's indignation was not unfounded. The Hungarian Communist Party (MKP), most likely at soviet pressure, did not use the Kovács-Nagy affair to cast discredit on the entire Smallholders' Party (KGP) or claim it was so riddled with reactionary elements that it must yield power to a party that stood closer to the masses, such as the Social Democratic Party (SDP). Nationalization of large enterprises was still proceeding slowly and the press remained fairly free. Speaking in the National Assembly, Rákosi rejected the charge that there had been a putsch and he was seconded by SDP speakers. On June 17 even the KGP deputies confirmed that the new government had been chosen according to constitutional forms.[8]

Meanwhile the peace treaty remained unratified. The cabinet had adopted a legislative bill endorsing it on May 16, but it was not brought before the National Assembly until after the new government had taken power. The foreign policy committee discussed it on June 16. One speaker, Dezső Sulyok, head of the recently formed Freedom Party, expressed concern, not about the treaty's international implications but about the leeway it might give the government in internal affairs, especially for the liquidation of parties with whose political orientation it did not agree. The committee thereupon adopted a resolution that expressly denied the government such freedom of action.

When the treaty came before the full assembly, Sulyok spoke again. He began with the leftist argument that it had been the sins of the old ruling classes that were responsible for the harsh treatment Hungary received at Paris, then he assailed the great powers (not excepting the Soviet Union) for giving Hungary not a peace of justice but a peace of punishment and vengeance.[9] He went on to state that Hungary should not make the business of the great powers its own; it must never again place its foreign policy in the service of another power or a group of powers. (This was a veiled reiteration of his previous proposal that Hungary pledge itself to be permanently neutral.)

Other critical comments were also voiced in the debate. The economic provisions of the treaty were unclear and would certainly not encourage speedy recovery. A number of speakers were chagrined by the fact that the treaty made minority protection a domestic issue; thus Hungarian minorities did not enjoy the protection of international agencies. What the debate showed was that a good part of the political right was not intimidated by recent events and remained deeply opposed to the treaty as a whole; the left on its part was ready to drop all objections to it.[10]

There was of course an air of unreality about the whole debate because the domestic and foreign relations of small states were increasingly being influenced by the global confrontation between east and west that left no aspect of public life unaffected. Stalin, through the pronouncements of his propaganda chief Andrey Zhdanov, had used the Truman Doctrine to produce a deep ideological chasm. The world by this model was divided into two camps, the "democratic" (for example, socialistic) and the capitalist-imperialistic; uncommitted nations, whether or not they were so inclined, had to choose between the two. There was no Third World as yet, but competition among great powers for the favors of nations emerging from colonial status had begun.

Szegedy-Maszák's temporary successor in Washington as chargé, Pál Marik, sent to the foreign ministry in Budapest a report of a confidential analysis that Generals George Marshall and Dwight D. Eisenhower gave members of Congress on international developments—though he did not identify his source. According to the report the generals had sketched an alarming scenario. Moscow allegedly had a timetable to create, before next summer, a Central European bloc to include Hungary, Czechoslovakia, Poland, Austria, Northern Italy, and Germany east of the Elbe. Soviet agents were reportedly active all over the world seeking to undermine the social fabric of noncommunist states. The prospect of Europe saving itself from these designs through its own efforts seemed practically nil. Governments that had so far escaped soviet tutelage felt they only had limited time left and, "in order to enable democratic forces to survive they must have almost unlimited help from the United States."[11]

The adoption of the Two Camps concept gave the communists in Hungary a powerful boost. Their struggle with reaction had at last received Moscow's backing. Democracy on the western model had lasted long enough. The free elections of November 1945 had provided an object lesson of how a revolutionary party could be swamped at the polls if it did not employ potent means of persuasion. Now the MKP called for new elections to be held as early as July.

In denouncing reaction the party could now cite the examples not only of amorphous political groupings but also of individuals who had resigned their posts abroad or had emigrated in protest of Nagy's ouster. Nagy himself had gone to America and, shortly after his arrival, secured an interview with Dean Acheson, perhaps the most powerful supporter in the west of "reaction." Acheson, out of courtesy or out of a genuine need for guidance, asked the fallen premier what policy he advocated toward East Europe, and Hungary in particular. Nagy tried to be helpful but could offer only platitudes. Europe must not be allowed to be separated into hostile camps, he said. Majority parties had to resist the tyranny of the minority and the United States had to give them every encouragement it could.[12]

Nagy then tried to appoint himself spokesman for the East European peasant masses. He joined two other peasant leaders, G. M. Dimitrov of Bulgaria and Alexander Cretzianu of Rumania, both of whom had suffered the same fate he had. On the Fourth of July the three men raised their voices "on behalf of those who cannot speak." They appealed to the democratic peasant parties "which so far did not have [the] opportunity to associate themselves with our efforts, to join us in [the] creation of a democratic international peasant union and eventual realization of a United States of Europe with the ultimate aim of establishing worldwide democratic unity."[13]

Communist parties everywhere adopted the word "democracy" as a synonym for a soviet system of government. The MKP extended its struggle against "antidemocratic forces" to the right wing of the SDP and put before it the choice of either accepting the communist program or being branded reactionary. Splinter parties such as Dezsö Sulyok's Freedom Party faced extinction; the political police shadowed the party's speakers at rallies and made verbatim notes of the speeches they delivered. These speeches were often fraught with meanings and messages long anathematized by the MKP. For instance, one Freedom Party speaker at a provincial rally said that "we hold that politics is the most sacred thing after religion because it means concern for the fate of the nation. . . . We were [once] members of the Smallholders Party and knew it received a majority [in the elections] because the people expected it to carry out their will [and preserve a system] resting on the inviolability of private property and a social establishment based on Christian thought and religious morality."[14]

Probably the worst thing the holdouts from the old order could do at this juncture was to associate themselves with the Church and religion in general. The Christian religious establishment had already been marked out as the next target of communist propaganda and the struggle became increasingly bitter because of Cardinal József Mindszenty's unrelenting campaign

against the "democratic" forces. Had the Catholic Church had a more pragmatic spokesman the focus probably would have shifted away from the
always troublesome religious question to more practical and less emotive
social and economic problems. As it was, with pulpits across the country
resounding with condemnation of the political line that was gaining control, there was no escaping the bruising struggle between church and state.

One report of the political police had it that Sulyok, with other former
right-wing members of the KGP, had gone to Esztergom to visit Mindszenty,
apparently to ask him if he was considering running in the upcoming elections. Mindszenty reportedly declined, saying that he was opposed to an
ideological party. Sulyok afterward expressed the opinion that the cardinal
was not being candid and would probably form a party before the elections
to monopolize the opposition to the communists. For an ephemeral season
though the flamboyant Sulyok rather than Mindszenty became the hero of
the political right, including diplomats from the west. At a party given at
the British embassy he was lionized by all those present while the premier,
Dinnyés, was practically ignored. Sulyok himself admitted it was embarrassing.[15] But within days Sulyok's bravado was gone.

On May 19 he visited Chapin and told him that he planned to retire from
politics before the elections. "It was clear," Chapin reported "[that] he was
personally endangered, would attempt to flee Hungary." The Minister
added that about 80 percent of the membership of the Freedom Party
had been arrested and interned in the past month, under the same law
that had put Sulyok into prison under the Nazis. "He pleaded [with me]
for assistance," Chapin went on, "in crossing the border with his family
. . . a request I have also received from other noncommunist groups. . . .
I should like to afford some assistance, [but] there appears to be no adequate and safe means at my disposal."[16]

Noncommunists also appealed to Chapin for funds in the election campaign. He always gave the correct answer that the American government
did not meddle in the internal politics of other nations. He drew the
counterargument that such an attitude was correct in normal times but "the
present situation in Hungary calls for special measures and that in line with
U.S. policy as announced in the Truman Doctrine, every possible step, orthodox or unorthodox, should be taken to counter intense Communist electoral propaganda."[17]

In this transitional period in Hungary's postwar history the conduct of
foreign affairs was entrusted to men of the most limited diplomatic experience; Mihályffi attended to the job only part time and the government was
in no hurry to appoint replacements for the envoys who had resigned be-

cause of the Nagy affair. (Szegedy-Maszák's successor in Washington did not occupy his post until September 30.) The diplomats who had gone into emigration exhibited, according to a prestigious Vienna journal, "the greatest self-restraint and self-discipline." In their pronouncements they supported a coalition government, albeit not the counterfeit one that emerged after Nagy's resignation. They were opposed to reactionary politics or any move that had the potential of causing conflict with the USSR. At the same time they were passsionately opposed to any attempt to establish the dictatorship of one party.[18]

By promulgating his doctrine Truman opened himself up to charges from sundry quarters that instead of lessening the tensions of the Cold War he had intensified them by turning a bogus emergency in the Eastern Mediterranean into a global confrontation. And so when George Marshall returned from another futile meeting of the Council of Foreign Ministers and brought with him appalling reports of economic destitution in Europe, a plan emerged from the depths of the State Department that the political damage done by the Truman Doctrine should be salvaged by a broad and nondiscriminatory offer of economic aid. Apart from its undeniably altruistic side the offer also reflected the conviction that economic breakdown all too often led to political radicalism in Europe.

The occasion was truly unprecedented. Only seven years earlier the United States had pronounced itself the arsenal of democracy and had undertaken to arm the forces fighting Nazism. Now it in effect offered to be the storehouse for a starved Europe. Hungary had been begging for American economic aid since the end of the war and while there had been some political intent in making the United States the nation's benefactor at a time when the soviets were plundering it, the aid was also crucial for rescuing Hungary from economic collapse. Washington, in considering these pleas, always kept an eye open for internal developments and sought to use economic means as reward or punishment for political behavior. (For example the Export-Import Bank, which had extended Hungary $10 million in cotton credits, canceled this loan because of the political changes following Nagy's resignation and because of the threat of the nationalization of the big banks.) Now there opened up the prospect of a vast infusion of material help with no visible strings attached. The offer was open to all European states, the USSR included.

What came to be known as the Marshall Plan caught Stalin off guard. A year earlier he had rejected an American loan offer of $1 billion (for which he had been pressing since the closing phase of the war), because the political price was too high (in effect a retreat from East Central Europe). Now

Washington demanded nothing in return for its proffered largesse. Stalin allowed Molotov to participate in a preliminary session of foreign ministers in Paris to frame a reply to Marshall's offer. But shortly after this conference opened he had a change of heart. He particularly objected to the American condition that the nations applying for the aid make their economic shortfalls public and try to overcome them by trading with one another before they presented the bill for the leftover needs to the United States. Molotov was ordered home and Stalin forbade the nations in the soviet bloc to avail themselves of the offer.

The soviet veto took different forms in different countries. Czechoslovakia received a peremptory demand while Klement Gottwald and Jan Masaryk were in Moscow. The day before they arrived their ambassador in Washington, Juraj Slavik, had received a communication from Prague stating that the Czechoslovak government would participate in the Paris conference. On July 11, in Moscow, the visiting statesmen confronted a different reality. Slavik heard of the change indirectly, from the United Press. He at once sought out the head of the Central European Affairs Office in the State Department to inform him that the offer he had made some days earlier no longer stood. His vis-à-vis remarked that "the decision not to participate, if it were correct, would certainly be interpreted as Soviet dictation of Czechoslovak foreign policy."[19] Slavik could only agree with that assessment.*

Gyula Szekfü observed the visit of the Czech statesmen closely. He reported home that after spending six days in the soviet capital Gottwald and Masaryk unexpectedly announced that they would be flying home at 4 A.M. the next day. The embassy explained the rather unusual hour with the statesmen's desire to discourage diplomats and the press from coming to the airport. In Szekfü's opinion Gottwald and Masaryk wanted to avoid any impression of having become soviet satellites.[20]

The Hungarian cabinet took up the Marshall aid offer on the morning of July 11. The meeting was chaired by Premier Dinnyés; he read a memo he had received from the soviet envoy, Georgy Pushkin. It stated that the soviet government had had little faith in the plan from the start, in part because it knew that England and France had come to an agreement with the United States behind the USSR's back and in part because the conditions of

---

*The Hungarian legation in London reported on July 23 the current view that Stalin had prevailed on the Czech statesmen to forgo Marshall aid by threatening that if they accepted it, he would support Slovak separatism, as well as the Polish demand for the region of Tesin. He would also move three soviet divisions to East Germany through Czechoslovakia.—New Hungarian Central Archives (NHCA), Budapest, Hungary, Foreign Ministry Files, English, Administrative, Box 9.

the offer were not sufficiently detailed. The soviet government perceived in the request that each government provide information about its needs an attempt to interfere in domestic affairs and impose on the countries an economic scheme that would complicate marketing their surpluses and make their economic lives dependent on American interests. With the views of the Soviet Union and the western powers so far apart, there was no prospect of agreement and the soviet government deemed it necessary to inform the Hungarian government of this, especially because a new conference on the Marshall Plan had been called for to open in Paris on July 12.

Acting foreign minister Mihályffi announced that Rumania, Bulgaria, and Yugoslavia had already rejected the invitations and that in Budapest the four coalition parties had held a conference and decided that the government should not send a delegation to Paris. Thus, seeing that the great powers were divided over the Marshall Plan and Hungary's neighbors refused to participate, he invited the cabinet to act likewise.

There was no hurry to concur with this request. Dissenting voices depicted the Marshall Plan as a purely economic offer that should be considered as such. Finance Minister Miklós Nyárády challenged the statement that Hungary's neighbors would not go to Paris; he contended that Austria and Czechoslovakia would (Prague's negative decision had not yet been made public). The minister of public works and reconstruction pointedly remarked that: "Reference to attitude of neighbor states shows the compulsion [by the USSR] of which it is needless to speak." The minister of agriculture gave it as his personal opinion that it would be unwise to stay away from the Paris conference and added that this was the public's view too. Given the country's economic condition he held it "suicidal not to grasp the last straw offered to us." He observed that in Italy for instance things had greatly improved since the country had started to receive American aid.

It was time for Rákosi to speak. He called his colleagues' attention to the fact that the interparty decision to refuse the invitation had been unanimous and he stated that he had nothing to add to it. But then he added a good deal. Public opinion could not be ascertained, he said, as no poll had been taken. Admittedly there were conservative circles who were sympathetic to the Anglo-American position and sought to push Hungary in the wrong direction. Thus references to public opinion could be ignored. The public had to be educated. It was gratuitous to say that the Marshall Plan was a last straw; that might have been true two years earlier when Hungary was like a drowning man, but no longer. Italy was in worse shape than Hungary even though it had been receiving American aid for the past three years. Hungary should not be intimidated by American threats. It was still receiving goods from the United States as part of a loan Washington had extended two years earlier.

The minister of agriculture objected that when he spoke of public opinion he was not referring to a small Budapest circle but to the villages. And to the entire Smallholders' Party. But Dinnyés, himself a Smallholder, disagreed. He had information, he said, that there was no financial backing behind the Marshall Plan. The U.S. Senate had not made a decision about it. He, Dinnyés, took personal responsibility for refusal.

Nyárády then came to the heart of the matter. The invitation to Paris, he said, had an economic and a political side. The latter had been settled by Pushkin's memo. But that didn't mean that the offer had to be refused. The government should at least express interest so that later it could not be charged with having been offered help only to cut it off at the source. The government could not ignore public opinion.[21]

Rákosi spoke, then sat quietly and listened. The argument was really not even over whether Marshall aid should or should not be accepted but over whether a delegation should be sent to Paris. Everybody knew that even if one was sent, the first step would be taken on a road with twists and turns no one could foresee. Those present could not but recall that a year earlier a government delegation had gone to the west against Pushkin's advice and although this act drew Stalin's displeasure, nothing worse was forthcoming. Since then Nagy and Gyöngyösi had been removed, the KGP had been splintered, as had the SDP, and the MKP secretariat had become the effective government of the nation. There was also the fact that a broad trade agreement with the USSR had tied the economic life of the two nations together by many threads. If American aid began to flow it would be difficult to stop, and any attempt to do so would produce a public outcry.

None of this was said because the meeting had been called to deal with the economic side of Marshall's offer. The cabinet in the end voted to send regrets to Washington. The minutes of the meeting record the text of the note to be sent:

> On the 4th of this month the invitation to the July 12 conference in Paris had been thoroughly examined by the Hungarian Council of Ministers which decided that while it greatly appreciates the significance of the efforts to rebuild Europe, to its great regret it sees no possibility of being represented at the Paris conference. It has become clear that in this matter the great powers do not agree. Hungary cannot participate in a conference over which the great powers are not in unison. With this decision the Hungarian government does not wish to exclude itself from the great effort of European recon-

struction and nothing stands farther from it than the rejection of the principle of mutual aid.[22]

When Chapin wrote to the State Department 12 days later he made no reference to the Marshall Plan but expressed alarm over "the rapid and grave deterioration in the Hungarian political situation." It was clear, he asserted, "that the Soviet Government is determined . . . to bring Hungary under complete domination and to incorporate it into the general Soviet system." He qualified this in a later paragraph: "I do not mean to imply that the Soviets intend immediately to force upon Hungary a complete Sovietization. . . . Most thinking Hungarians with whom I have talked agree that for the time being this is a secondary objective and that the Soviets and their Hungarian Communist allies are far too clever to arouse unnecessary opposition at this stage by enforcing collectivization . . . or by direct attacks upon Magyar culture or religion."[23]

The contradictions in the report illustrated its alarmist tenor. Moscow's determination to bring Hungary under complete soviet domination did not dovetail with its concern over arousing "unnecessary opposition" or its reluctance to enforce collectivization. The fact was that "at this stage," apart from the judicial fallout from the Kovács-Nagy affair, there was no evidence of any deliberate drive to sovietize Hungary. The press was free and debate in the parliament and its committees were untrammeled. The border to the west was open. The great bulk of medium and small business remained in private hands. But Chapin, like envoys in other soviet-bloc countries, was guided by preconceptions. To permit Hungary "to regain its independence," he wrote elsewhere in his report, "would split the satellite Slav states and interfere seriously with the [soviet] plan for domination of South Eastern Europe." There were only two satellite Slav states at this point, Poland and Bulgaria, and they did not form a bloc that an independent Hungary could split. Possibly Chapin wrote what he assumed the State Department wanted to hear to underpin its general policy.

But such opinions were current in Great Britain too. His Majesty's Government (HMG) generally maintained a deep diplomatic silence in its relations with Hungary but certain press organs, notably the *Economist*, often echoed the views of the government. It theorized in July that the soviets intended, before their withdrawal from East Europe, to cleanse the governments in the region of all pro-American elements. Fear was growing that the Hungarian elections, scheduled for August, would be far less free than the previous one had been. Nagy was purged precisely because he based his program on the expectation of American support—thus the farcical situation arose that a premier was accused of conspiring against his own government.

The western powers were apprehensive that the USSR sought to create a solid bloc of satellite states from the Baltic to the Balkans before American aid produced an economically united Europe.[24]

In Hungary there was evidence to the contrary and the leftward pressures were often more psychological than political. Sulyok's Freedom Party dissolved probably less because of MKP harassment than because its founder was afraid to enter an election campaign that promised to be rough. Another dissident Smallholder politician, Zoltán Pfeiffer, formed a new party from fugitive members of the Freedom Party. Referring to the communists as the "counter party," he accused them of undermining the KGP and even appointing its leaders. He pledged "to unfurl the banner of a truly independent party," the "moral foundation" of the old KGP.[25]

Preparations for the elections confirmed some fears. They were marked by the disenfranchisement of a great many voters and the enablement of professional communists and members of the political police to cast multiple ballots in different places. London was more concerned than Washington was; the Hungarian envoy reported that Hungary's political balance sheet in England was shifting from the credit to the debit side. He made some cynical comments in this connection. "Possibly English society puts more emphasis on clean elections . . . than we anticipated. . . . Perhaps many in Hungary will shrug their shoulders and ask whether England does not have other things to worry about than whether a few Jews returned from concentration camps will be given the vote* or whether a few persons with doctorates end up disenfranchised as mentally deficient."[26]

Hungarian visitors to England offered explanations to official and semi-official personages of why the elections should be restricted. The masses were politically immature, they contended, and "must still be protected from the dangers of their political instincts." The reaction to such arguments in Anglo-Saxon countries was that they had heard enough of the immaturity of the Hungarian masses. It was probably this attitude that had drawn Hungary into another losing war.[27]

Altogether seven nonleftist parties participated in the elections. This fragmentation enabled the MKP to emerge as the single largest party with 22.25 percent of the vote, an increase, though not a dramatic one, from the 17 percent the party got in 1945. The KGP vote fell to 15.38 percent of the total but it was clear to everyone that many of the votes cast for smaller parties would have gone to the KGP had the opposition to the communists not been splintered. The SDP got 14.88 percent and the National Peasant

---

*A number of such Jews were denied the vote on the ground that during the war they had gone to the west.

Party (PP) 8.3 percent. Altogether the left bloc still had less than half of all the votes cast whereas the bourgeois parties had 55 percent.

Irregular practices aside, the elections were orderly and free of intimidation. But in a larger sense they were of no serious consequence. The political future of the country depended on whether the parties on either side would continue to exercise the restraint and practical good sense they had exhibited thus far. It also depended on whether the great powers could prevent polarization of the world into hostile camps.

Chapter 17

Crossroads

On August 13, 1947, as an election campaign fraught with dubious practices was in full swing, Hungary concluded with Great Britain a commercial treaty calling for Hungarian food shipments in exchange for capital machinery. The treaty came on the heels of failed talks of a similar nature between Great Britain and the Soviet Union. Moscow experienced the first unhappy consequences of its veto on satellite participation in the Marshall Plan. By forbidding the border states to receive American grants, the soviets accepted implicit responsibility for providing them with goods the United States would otherwise have provided. The result was a reverse pattern in established soviet commercial practice: the USSR was obliged to export whereas till now the flow of goods had been in its direction only. When the talks with England got under way it became clear that the British expected heavy grain shipments from Russia (at a time when the soviet people were near starvation). Moscow, unwilling to admit that its available grain would henceforth go to East Europe, demanded impossibly high prices for it and the British terminated the talks.[1]

The Anglo-Hungarian trade agreement did not have an easy time in the House of Commons. Robert Vansittart, an old and respected member of parliament, read a letter Ferenc Nagy had written after his resignation that spoke of a soviet "master plan to control East Europe." This revelation elicited a comment from the benches to the effect that while His Majesty's

Government (HMG) was desirous to trade with Hungary, "the result of the present situation is that Hungary will become a totalitarian state, these visions will fade and we must necessarily revise our attitude."[2]

Such a negative view in capitalistic states could put Hungary's international position in serious jeopardy. In the past two years the struggling country had placed its relations with foreign states (Czechoslovakia being a notable exception) on a sound and confident footing. The land reform, the proclamation of the republic, and the nation's emergence from an unprecedented economic crisis had created genuine sympathy abroad. Now goodwill began to erode. The western powers found that they were dealing with a schizophrenic political entity, divided between prowestern public sentiment on the one hand and a prosoviet ruling elite on the other.

It was just at this time that, with the ratification of the peace treaty, Hungary regained its full sovereignty. On September 15, the Allied Control Commission (ACC), hitherto an instrument of soviet control, went out of existence. The bulk of soviet forces was due to be withdrawn shortly. These items figured on the positive side. On the negative side was the fact that there would still remain a more than symbolic presence of soviet troops and that the political right had become so divided, and a substantial portion of it so intimidated, that it could no longer be depended on to be a counterweight to soviet influence. The premier, although a Smallholder, at times outdid the Hungarian Communist Party (MKP) in supporting soviet desiderata. Lajos Dinnyés had in effect become Mátyás Rákosi's parliamentary clerk. When he received a letter from Zoltán Pfeiffer complaining of police harassment of Pfeiffer's National Independence Party, instead of acting on it, he sent it to Rákosi "for your kind consideration." Rákosi left the letter unanswered, and so did of course Dinnyés himself. Thereupon other members of the party, which had become a minor thorn in the side of the MKP, wrote to Mihály Farkas, who routinely handled public relations matters for Rákosi, requesting negotiations with "the leading party of the coalition," the MKP, so that they could gain a clear picture of their party's position. Farkas did not answer, an associate of his did, saying that the MKP had nothing to talk about with members of the National Independence Party, there had to be a misunderstanding.[3]

József Mindszenty, irrepressible, read the signs correctly and, bypassing the premier, sent his letters directly to Rákosi. In June he registered a bitter complaint against a new marriage decree that dispensed with a church ceremony; it was, the cardinal asserted, an attack on the sanctity of marriage. He also reminded Rákosi that the Church was being denied a daily press organ and that even its biweekly publication did not receive the paper it needed, that Church buildings were being expropriated for the use of communist youth organizations, and that in civil service appointments Catholics were being discriminated against. "Unwarranted advantage [is being given] to Israelites and Protestants in violation of democratic principles."[4] General George Weems, the U.S. member of the defunct ACC, in his summary report on his tenure, placed the first Hungarian political crisis in July 1946 when Vladimir Sviridov ordered the government to purge itself of antisoviet elements. Coercion had continued ever since and had gained new impetus after the Kovács-Nagy affair. Courts were handing down sentences of long prison terms and even death. The government of Dinnyés, "a malleable Smallholder," had issued a White Book to document the existence of a conspiracy but offered no conclusive proof.[5] While Weems judged the hold of the MKP as enormously strengthened, he felt that "the way is still clear, should the US Govt determine that the effort is worthwhile, to establish effective bonds between Hungary and the western powers under the terms of the Treaty of Peace which comes into force today."[6]

Neither of the Anglo-Saxon powers apparently deemed the effort worthwhile. When at the end of the month the United Nations (UN) Security Council debated the admission of Hungary, as well as the other defeated states, to the UN, the United States and Great Britain abstained. As the bylaws required the positive concurrence of all the great powers, admission was denied. The government in Budapest angrily denounced the stated reason for the refusal, that Hungary was not able to fulfill its obligations under the UN Charter.

Nevertheless, when Dinnyés outlined his program to the National Assembly on October 7, he laid special stress on the need for good relations with the United States, England, and France and expressed himself dismayed over the fact that on the basis of "false information and allegations" Hungary had been denied membership in the UN. "Hungarian democracy," he said, "in full possession of its sovereignty, will take special care that there should be no interference from outside in her internal affairs and if it does happen we will, as in the past, so also in the future, reject every such attempt most energetically."[7]

It was their ritualistic tenor that deprived such pronouncements of any substance. Still, when József Révai took the floor, he criticized Dinnyés's program on different grounds. He denounced the notion (which Dinnyés didn't really express but which was the cornerstone of the Smallholders' program) that because Hungary was a small state she should stay away from global politics and let the great powers fight among themselves—to his mind this was just another version of dishonorable neutrality. The MKP, he announced, would be glad if the great powers resolved their differences, but if there was to be an agreement among them, it should be such "that it should have as its outcome the bankruptcy of imperialist designs." He rejected the very word neutrality. Behind it, he said, was the suggestion that Hungary looked on the great powers as equal to one another and regarded Anglo-Saxon aspirations as "power politics of the same kind as those aspirations which in international politics are first and foremost represented by the Soviet Union. We communists are not inclined to make such an equation. We contend that there is on the one side the country of socialism and on the other monopoly capitalism and its allies."[8]

To be sure, this was the position of only one party among many, albeit the one which had received the most votes. And although Révai's voice was Rákosi's voice and Rákosi spoke for Stalin, Hungarian foreign policy was still directed by men holding less extremist views. Ernö Mihályffi was a cipher but the officialdom at the foreign ministry had been reared under János Gyöngyösi's tenure and most of the ambassadorial posts were filled by diplomats of the old guard. Gyula Szekfü was lecturing the foreign ministry in Budapest about the need to defer to Moscow's wishes but his entire political past belied that urging. And Szegedy-Maszák's successor in Washington, presenting his credentials just at this time, was of the old guard in more ways than one.

At age 75, Rusztem Vámbéry looked back on a rich political and cultural life as a journalist, law professor, and publicist, though his ideological convictions were obscure. Not a member of any party, he had been loosely associated with the Social Democratic Party (SDP); he had lived in the United States for years. He was uneasily trying to find his ground between a Marxist world view and the democratic traditions of the United States. In meeting Assistant Secretary of State Frederick Merrill, he struck a buoyant, optimistic note. He portrayed Hungary as brimming with hope and cheer, much more so than other European nations, "including the Czechs." Much of the dynamism, he said, stemmed from the exertions of the MKP, which was "reasonably independent from Moscow's dictation." He made light of rumors of election fraud and denied that the Nagy affair had been a communist coup. Warming to his subject, he confided that Georgy Pushkin had told him that "the USSR could have had Hungary in 1945 if it had so wished. But Moscow wanted Hungary independent, so it could trade with the West and rehabilitate its economy so that in the end it might be exploited more profitably by the USSR."[9]

Possibly in a different international climate this statement, from the mouth of a diplomat not given to extravagant claims, would have occasioned deep reflection in the State Department. A sober review of events in Hungary since 1945 might have confirmed the validity of Pushkin's claim. But the mindset in Washington was no longer receptive to such departures from Cold War mentality. Even Selden Chapin, a confirmed pessimist, expressed surprise in one of his reports that "this process [of sovietization] does not proceed more rapidly for certainly there is no organized political force within the country which can stand in its way."[10]

He arrived at a somewhat romantic explanation. There was, as he saw it, amid the political misdirection, a hard vein of nationalism that "has survived Turkish, Slavic and German overlordship." In his view, "it is this sense of racial and cultural survival traditionally bound up with a feeling of pride at having served as the easternmost projection of Western civilization which differentiates Hungary so much from the surrounding satellites."[11]

Most Hungarians in fact liked to think of their nation as this easternmost projection, and hence the outpost, of a civilization that was now abandoning them to still another onslaught from the east. They had over the centuries adjusted to a communality with German-speaking peoples, western, Christian, and culturally enlightened. But any affinity with the Slavs was totally alien to the Hungarian temper and recent contact with the Red Army had powerfully fortified that bias. If there was any truth in Pushkin's version of soviet strategy, Stalin, at first, instead of combating the antipathy of Hungarians to anything Slavic, tried to isolate it, by making common cause with the Slavophile elements in neighboring states. The MKP's task apparently, was not to be the vanguard of sovietization, but merely to act as a brake on the impulse for a western orientation. But the reaction proved much stronger than Stalin had reckoned and it did not appear feasible to allow Hungary independence and at the same time curb its ties with the west.

The analogy with Finland did not work because Finland had nothing that would compare with Hungary's political right. Also, Finland had been a *cas de conscience* with the western powers since the Winter War of 1939–40 when they allowed it to be defeated by the Soviet Union; Hungary had no redeeming feature that would endear it to western hearts. On a more practical plane, the Finns were faithfully carrying out their reparations obligations to the soviets,[12] whereas Hungary was forever in arrears and seeking western support for the reduction of the reparations bill. In Finland, the political scene was quiescent; in Hungary it was an ideological battleground. Finally, Finland was not an essential land bridge to other states in which the USSR had vital interests.

The Two Camps concept had by now in any case made an uncommitted country an embarrassing anomaly. The USSR had been given a 75 percent influence in Hungary; in the "legal" sense that settled the matter. Still, Stalin proceeded with caution. The Americans showed a disconcerting interest in Hungary, yet American disinterestedness was essential for Stalin to pursue his foreign policy objectives in Europe. There were indeed two camps, but Stalin lorded it over a backward and disinherited domain, even in Germany, where as things were shaping up in that once powerful country, the rich western two-thirds was on its way to becoming united (on January 1, 1947 the British and American occupation zones had been combined and the pressure was increasing on France to add its zone and create Trizonia) and economically viable. The trend could only be checked by military means (the one area in which the soviets had the advantage) and Stalin had excluded that means as an instrument of national policy.

The kind of "bridge" Stalin envisioned Hungary might become would, in a stalemate, serve as a link between the prosperous and the impoverished camps. It was a pity that the Marshall Plan was such a flagrantly political scheme. Stalin did not oppose American material aid to Hungary, under UNRRA auspices or as loans or credit for purchases of surplus—as long as it was not part of an embracing effort to make Europe dependent on an infusion of American aid. Possibly, had the cabinet decided to accept Marshall aid, soviet control would have been tightened. But after the rejection of the offer pressure remained mild and the general atmosphere was nowhere as depressing as diplomatic reports usually described it. József Révai was not far from the truth when, speaking to an American journalist who inquired about the "terror" in Hungary, he said that such a contention was at variance with facts. "If as a seasoned reporter," he said, "you walk the streets of our capital for only an hour, you find the answer yourself. No guards patrol the streets, traffic policemen are unarmed, you are not asked to identify yourself, not even if you travel from one end of the country to the other. Hungarian democracy employs 28,000 policemen and 12,000 soldiers. No other country in Europe has fewer. Our strength lies in our democratic workers and peasants."[13]

This was the voice of confidence; the MKP was still hoping that once the forces of reaction had been overcome the public would readily follow it on the road to socialism. Stalin appeared to be willing to settle for even less, a limited private enterprise with only the "commanding heights" of heavy industry in government hands, and even those open to accommodation with western interests. Chapin had remarked on the slow pace of "sovietization"; Weems at the same time saw a determined effort to achieve just that, arguing that the Soviet Union could not afford to "split Slavic unity" by tolerating an independent Hungary. But Stalin had a solution for avoiding such a split without outright sovietization. It wasn't so much ideological dissonance that concerned him but the age-old animosities in the Carpathian sphere. Thus, in the closing months and weeks of 1947, at Moscow's behest, a number of friendship treaties were concluded among the people's democracies in the region.

Hungary joined this concord late, both because of its special staus and because most of those animosities centered around it. To embrace as friends neighbors upon whom the public looked as despoilers of historic lands required a massive propaganda campaign. As early as September 12 *Szabad Nép* took pains to rebut the contention that the neighboring states, by drawing together, were seeking to encircle Hungary and create a new Little Entente. On the contrary, the paper wrote, the purpose of the treaties (which at that point were still being negotiated) was to ensure the cooperation of peace-loving states. Hungary could not stand apart and become a springboard for western imperialism.[14]

The obligatory references to Marxist solidarity were conspicuously absent from these pronouncements. It was indeed not clear just what part the communist parties should play in the process. Except perhaps in Yugoslavia, party politics and national politics had not yet fused. Indeed a disorientation as to the proper conduct in national politics had become endemic in communist parties, inside the soviet bloc and beyond it. There was a danger that communist parties, distracted by the exigencies of national politics, would lose their vision and become mere players on the scene instead of vanguards to a brave new world.

And so, in September 1947, the communist parties of the soviet bloc, as well as those of France and Italy, were summoned to a conference to be held in Warsaw, in order to hear from the lips of the soviet delegates what they had done wrong and what they had done right and how international unity could be strengthened. The MKP sent Révai and Farkas; the report they prepared upon their return was for communist eyes only. That the Warsaw conference was of the greatest importance to the soviets they gathered from the fact that Stalin had sent two of his closest associates, Georgy Malenkov and Andrey Zhdanov, to be his spokesmen.

The report makes it clear that there was at the meeting as much dissonance as harmony. Some of the disagreements centered on the dispute between Hungary and Czechoslovakia, which was discussed at great length and elicited much angry rhetoric. The Czechoslovak representative, Rudolf Slansky, accused the Hungarian comrades of sabotaging the peace treaty that had ordered the two sides to reach an agreement about the fate of Hungarians in Slovakia who were not included in the population exchange. As a matter of fact, Slansky charged, Hungary was not even carrying out its part of the exchange. For a thousand years Hungarians had dominated Slovaks and forcibly Magyarized them. As long as there were Hungarians in Slovakia, revisionism remained a threat.

This was hardly the kind of talk suited for a gathering of communists with internationalist views. Slansky defended it by reminding his listeners that at the peace conference Andrey Vishinsky had supported his government's position; consequently it had to be right. He claimed that documents showed that Hungarian communists in Slovakia had worked against the Czechoslovak state; the truth was, they did not understand the raison d'être of a Czechoslovak state and strove for domination.

But this position received no support from the comrades. The soviet delegate did not comment on Slansky's charges but Anna Pauker, speaking for the Rumanian party, Eduard Kardelj, representing Yugoslavia, and Vlko Chervenkov, from Bulgaria all spoke against it. This however was only a distraction from the main subject, which was the examination of the policies and methods employed by each of the communist parties represented at the conference. Malenkov, outlining conditions in his native USSR, explained how the transition from war to peace had complicated the work of the Communist Party of the Soviet Union (CPSU) and placed the struggle against American imperialism in first place among its tasks. "In the Soviet Union," he said, "international imperialism can no longer rely on [reactionary] class elements, therefore it seeks to rely on tendencies that are equivalent to bowing before bourgeois culture. The intellectuals [guilty of this] easily become victims of spy organizations."

Kardelj spoke next and announced that the Yugoslav party had largely liquidated reaction. It did not tolerate organized opposition to the party line "at a time when the economic superiority of American imperialism . . . is so immense. An opposition party [ipso facto] becomes a tool of American imperialism."

The soviet and Yugoslav delegates, who spoke with the voice of authority, judged the other parties represented at the meeting deficient for one reason or another. Malenkov and Zhdanov saved their worst accusations for the French and the Italians. The delegates from those two countries defended themselves by claiming that their governments had unconditionally bowed to American economic pressure. The French delegate, Jacques Duclos, declared that his country was in danger of becoming an American satellite. "We would be much stronger economically if the government's policy were different. But many government contracts which could be filled by French industry, were given to Americans." The situation in agriculture was catastrophic, he further reported. For that too the United States was to blame. Twelve years earlier five million hectares in France had been planted with grain, in 1947 only 3.3 million. The reason was that the government fixed prices so that farmers had no incentive to grow grain. It did this in the interest of the United States. At an agricultural conference held in Copenhagen the American delegate had declared that France should grow more flowers and less wheat.

Luigi Longo, speaking for the Italian party, voiced similar complaints, but Zhdanov repeatedly interrupted him. He demanded to know why the Italian Communist Party allowed itself to be excluded from the government. Did it have plans to return to the cabinet? The CPSU was willing to condone the fact that the Italian comrades did not want to involve themselves in adventures, Zhdanov said, but would they call strikes and demonstrations adventures? In general, the French and Italian parties were declared guilty of inaction, of failure to expose the true aim of the Marshall Plan, which was to place Europe into the shackles of colonialism. The French communists were remiss also in not exposing the maneuverings of the Social Democrats who had betrayed the cause. "You have allowed yourself," Zhdanov charged, "to be taken in by talk about France's national interests. You have allowed yourself to be blinded by accusations that you are not sufficiently patriotic." Yet, he stated, the only truly patriotic force in France was the Communist Party.

The conference established, in place of the defunct Comintern, an information bureau, informally referred to as the Cominform. It was not however intended to be a substitute for the former organization. It was not global in scope and membership in it was voluntary. Its main task was to redirect the struggle against fascism to the struggle against imperialism. Although government coalitions by "democratic parties" were still necessary, the French and Italian examples had shown that coalitions were meaningful only if they were used for the acquisition of a power base for the communists. "The Yugoslav example should be our guide. The difference between the Yugoslav National Front and other people's democracies is that in Yugoslavia the commmunist party had absorbed and liquidated National Front forces, [while] in other countries [the coalition] is based on the principle of parity."[15]

It was perhaps in recognition of the Yugoslav party's achievement that the conference decided to place the headquarters of the newly formed Cominform in Belgrade. Yet already the seeds of deep soviet-Yugoslav disagreements had been sown and possibly Stalin intended the Cominform headquarters to be a listening post in Belgrade to report on new acts of Yugoslav disobedience.

In reporting to the MKP central leadership Révai and Farkas asserted that the course followed by the Hungarian party had on the whole been the right one, although there still were shortcomings to overcome. There had been much talk about the need to cooperate with parties in other countries yet little had been done in that direction. Also, the party had often been guided by "practicism" (pragmatism?) and had not paused to review its ideological position. Economic policy had to be revised with a view of squeezing the bourgeoisie out of the economic life of the country. The party's relations with the military and the trade unions also had to be reviewed.[16]

Commenting on the Warsaw Conference in an unpublished article, Rákosi wrote that it had been made necessary by the aggressive reactionary policies of the United States. He called attention to the fact that in World War II not all enemies of fascism had fought for the same goals. The USSR saw in fascism an enemy of social progress; the western powers saw it as a rival in world markets. Of the imperialist powers the United States alone came out of the war strengthened; it now sought to use its position to achieve world dominion. It wanted to oust Britain and France from their positions of world power as it had already done in Greece, Egypt, and the Far East. But the USSR and the people's democracies stood in the way. Now the United States tried to intimidate them with threats of a new war. It was due to this policy that there were two camps, and hence the need for the people's democracies to achieve unity.[17]

The friendship treaties in the soviet bloc that now followed in quick succession were at least partly in answer to the need for such unity. Nations with a long history of animosity at times bordering on hatred discovered their common ideals and interests. Hungary signed a treaty of friendship and cooperation with Yugoslavia on December 8, 1947. (A more limited cultural agreement had been concluded in October, the two countries agreeing to harmonize their cultural endeavors and establish institutions to study each other's cultures.)[18] The political treaty contained pledges of consultation in matters of peace, war, and international relations. Other such compacts followed. Relations with Rumania had improved, thanks largely to Petru Groza's tolerant nationality policy. But many obstacles to friendship remained. The policies framed in Bucharest were often vitiated on the local level. A number of interstate issues also had to be cleared away, notably the release of frozen Hungarian assets and the status of ethnic Hungarians who had been deprived of their citizenship. Mistrust still ran deep. Groza had nevertheless generated enough goodwill to allow a friendship treaty to be concluded and ratified with little difficulty.

Chapin, in a report of February 11, 1948, cited this treaty as proof that communist control over Hungary "has been further reinforced and tightened." He saw other evidences as well. The fusion of the MKP and the SDP had been all but consummated and the foreign ministry had undergone a sweeping purge; he also referred to a lecture given at the soviet legation to communist officials on tactics to defeat the Marshall Plan.[19]

The road was now clear for a treaty between Hungary and the Soviet Union. The new Hungarian foreign minister, Erik Molnár, advised Pushkin on January 27 that his government intended to send a delegation to the soviet capital for the conclusion of a mutual-aid treaty.[20] According to Pushkin's report the foreign minister also intimated that he hoped to gain soviet support for the creation of a Hungarian armed force.

This communication was preceded by a number of concessions by Moscow to the still-struggling Hungarian economy. The conclusion of a new commercial and maritime agreement was the occasion for the announcement that the soviets were returning to Hungary 381 steam locomotives and 3,836 railroad cars that they had seized as war booty. On December 20, 1947 Moscow had advised the Hungarian government of a great reduction of soviet military personnel on Hungarian soil. That same day Pushkin communicated the decision that the price of goods Hungary delivered in reparations would be raised and that such goods no longer had to be shipped to the Soviet Union but could be handed over in designated Danube ports. (According to press reports these concessions lowered reparations obligations by over $17 million.) Finally, at the end of January 1948, talks on the repatriation of prisoners of war were successfully completed.

Szekfü saw his admonitions of the year before vindicated. "I have said," he wrote to the foreign ministry, "that our future can be assured only if we accept satellite status as other states bordering the USSR have. For this we had to gain soviet trust and that we could do only if we put an end to the assumption that reactionary forces can yet return and turn back the clock of history." With a long reactionary past of his own, Szekfü bowed to the moral necessity of satisfying soviet expectations. "In the end," his report continued, "the nature of our relations with the Soviet Union is entirely a matter of trust. Examples of Finland and Rumania show that when trust is there, the soviets become malleable. . . . We have to continue movement from a capitalistic to a socialistic direction and pass from individualistic ways to a somewhat more collective world."[21]

The treaty was signed on February 28, by Dinnyés for Hungary and Molotov for the Soviet Union. Asserting that the pact served the interest of strengthening neighborly relations, the contracting parties undertook to oppose the danger posed by a resurgent Germany and to lend each other, in case of aggression by Germany or any nation allied with Germany, every military and other aid. The parties also pledged that they would consult with each other on all important matters and that neither would participate in an alliance directed against the other.[22]

Most noteworthy about the pact was that, unlike the ones concluded by the USSR with Yugoslavia and Rumania, which obligated the parties to help each other against any aggressor, this one referred only to Germany and its possible allies. On the other hand it was not purely defensive; it undertook offensive obligations as well. There also were pledges of noninterference in each other's internal affairs.

The ratification debate started on the very day the delegation returned from Moscow and lasted two days. Given the unanimity of support for it, the debate needed to take up only an hour or two. But the deputies were intent to enter their enthusiasm for the pact into the record. Gyöngyösi, now a deputy, expressed hope that the treaty would ease the lot of Hungarians in Czechoslovakia. Révai noted triumphantly that Hungary had finally made its choice between the two camps. He went so far as to declare what he could not help knowing was untrue, that unlike those among other nations, treaties between the Soviet Union and the people's democracies did not lead to the formation of blocs or the domination of one party by another. He acknowledged that the presence of soviet troops in Hungary influenced the country's internal politics, but added that the same held true for the presence of British troops in Greece—with a difference. In the former instance the influence worked for the benefit of the people whereas in the latter it worked against it.[23]

There was little doubt in the public mind that these accolades sought to conceal the still-live hostility among Hungarians against everything Russian. The restoration of the country's sovereignty had led to a sense of confidence, not only among the people but in official circles as well. For instance László Pesta, who as the chairman of the Reparations Bureau was responsible for the fulfillment of related obligations, made it clear in an internal memo that after the ratification of the peace treaty Hungary was no longer bound to defray the expenses of soviet occupying forces, either in money or in kind. As for the remaining liaison troops, Article 22 of the peace treaty held Hungary responsible for providing funds, goods, and services for their upkeep and transportation. Even that, however, was to be rendered only if the goods or services rendered were expressly for the purpose of facilitating liaison with Austria, if they were specifically requested, and if compensation was given. "We are a sovereign nation," the memo noted, "no longer obliged to defray costs of foreign troops. If the soviets feel differently, let them enter into conversations with us. Soviet forces have to evacuate hotels, private apartments and private institutions they have so far occupied and be put up in barracks, warehouses and garages specifically designated for that purpose."[24]

In the early weeks of 1948 it was still possible to prognosticate that, barring adverse international developments, the political balance in Hungary would continue. The right and center were fervently looking forward to the conclusion of peace with Germany because it would as a matter of course have been followed by an Austrian treaty and, with the occupation of that country ended, the soviets would no longer have had an excuse to keep troops in Hungary. That would have been the true restoration of the country's independence.

In the offices of the MKP central leadership a memorandum was prepared by an unidentified official that accurately revealed the uncertainty of the future. It was a strange mixture of realism and self-delusion:

> Our party came out of the 1947 elections strengthened and with the reaction pushed back. On the lower levels though the reaction took the offensive. It enlisted new masses into its ranks and infiltrated the working class, even the Communist Party. This temporarily clouded our achievements.
>
> In 1945 reaction had won a victory [in the elections] and democracy suffered a defeat. In the East and Southeast European state [structure] Hungary proved the weakest link. This was why foreign reaction targeted Hungary and domestic reaction was encouraged. Had we allowed the process to continue, we would now be in a semi-colonial status to the United States; instead of peaceful development, we would be plunged into a civil war.
>
> If we now move forward it is thanks to the struggles of the MKP. We kept Hungary independent. With us patriotism was not a parliamentary maneuver. . . . . What is now the situation? Reaction has no unified party, it will never have one, the front of the reaction is broken. We now have the presidency of the parliament, secretaries in important positions and we head the foreign ministry. This gives us a weapon in the struggle against international reaction.
>
> We cannot build democracy, let alone socialism, with forces which regard western democracy as their ideal. It is impossible to promote our growth with those who want to cooperate with the Soviet Union only superficially, or with such as do not desire the fusion of the two workers' parties. But for that to happen, the SDP must purge itself.
>
> What is the international situation? The warlike mood begins to abate. There is no danger of war. Why? Because America is weak. In vain has she grown economically by developing her industry and hoarding gold; for a new war she is weak. Also, a war against the USSR would

have to be fought on dry land and it is commonly known that the USSR is the strongest land power.

Why does the United States spread war propaganda then? Because she wants a number of European states to capitulate and accept the conditions of American imperialism. That would mean colonial subjection. But if we accept the battle, they will have to retreat. The crisis of American imperialism is inevitable. It will be a crisis such as the world had never seen. This is why they are intent on avoiding it.[25]

Chapter 18

Achilles' Heels: Minorities and Religion

Hungary had now formalized its friendship with all its neighbor states except Czechoslovakia; to fail making a similar treaty with it would have meant admitting the inadmissible, that differences between the two countries were still too great for a formal pact of friendship. The population exchange was proceeding extremely slowly. Of the 92,390 Slovaks who had registered for resettlement 73,273 had left the country. But only 68,207 of the 105,047 ethnic Magyars slated for transfer to the mother country had actually made the move. Another 6,000, not included in the formal exchange, had resettled voluntarily.[1] The discrepancy in figures was all the more puzzling as the Slovaks in Hungary were under no pressure to leave, nor were there any discriminatory measures instituted against them. At the same time the Magyars in Slovakia were being mercilessly harassed. The Hungarian legation in Prague was besieged by people seeking relief from the persecutions.[2]

And yet the two countries had a political feature in common that should have been an incentive to them to draw closer together: both sought to preserve a multiparty democratic system when all the other east-bloc countries had essentially succumbed to communist domination.

Neither Hungary nor Czechoslovakia was in a hurry to choose between the two camps, both paid lip service to the principles of people's democracies while trading liberally with the west, and culturally both had western preferences. The American envoy in Prague reported that he believed 80 percent of the Czech people favored western-style democracy over communism, "but expedience and timidity render most of them inarticulate."[3] He might have added that the same held true for Hungary, though there the opposition, now speaking in a lower voice, was still articulate.

Of the two countries Czechoslovakia appeared safer from sovietization. Free of occupation troops, it enjoyed wide international sympathy and a coup d'état would produce vehement repercussions. Yet, in two tumultuous weeks in February 1948, the government fell into communist hands. There had been faint signs that a change was coming. At the Cominform conference the Czechoslovak delegate had apologized for the slow pace in the "reorganization of the State apparatus," but explained that "Beneš and company," with their anti-German reputation, could not easily be removed from their leading position. In the same breath he criticized Eduard Beneš for being opposed to rapid nationalization (a course the communists favored).[4]

In the elections of May 1947, 38 percent of the vote had gone to the communists; since then, due to many political and economic mistakes, their popularity had sharply fallen. New elections were scheduled for May 1948 and it would have been extremely embarrassing to the soviets if the party had fared poorly. The communists had a formula to prevent this, one that had been successfully employed by Josip Broz Tito and lavishly praised at the Warsaw conference. The duty of communist parties in a National Front was, according to this formula, not to practice coalition politics but to undermine and ultimately absorb the opposition parties. The Communist Party, in its New Year message to the Czechoslovak nation, had put the position in a characteristically orotund fashion: "[T]he communists do not see the National Front as a coalition of political parties . . . it is an alliance of workers, peasants, handicraftsmen and intellectuals, an alliance which must become firmer every day."[5] The message was veiled but unmistakable: party labels no longer mattered; the Communist Party alone was ideologically capacious enough to provide a political home for all those who worked with their hands and their minds.

The issue that precipitated the crisis was a familiar one: communist control of the police. It was, to be sure, not of the magnitude to occasion a radical political change. In Hungary the year before this same problem had been painlessly solved and even a more serious one, a conspiracy case involving the premier, had been weathered without major damage to coalition politics. But in Czechoslovakia the Communist Party did not allow the opportunity to slip. Opposition parties were naive enough to believe that if their cabinet members resigned over the question of police control, the communists would back down. As it happened they did not and by February 24 they had taken charge of the government. The American ambassador, Laurence Steinhardt, thus far a staunch defender of the republic, was appalled. He charged that the communists, "by intimidation and demonstration of armed force have succeeded in seizing the government, eliminating all opposition." In a subsequent report, on April 30, he assessed the situation more calmly. The crisis, he wrote, almost certainly had been precipitated by President Beneš himself. Moreover, "the extent of soviet threats was probably less than on similar recent occasions in Finland and Iran, both of which countries successfully resisted such threats whereas the Czechs succumbed to them." He stressed that there was no direct evidence of soviet interference.[6] The pattern was nevertheless clear and predictably would be repeated in other states: the soviets allowed their domestic allies to do the messy operational work while leaving no doubt as to on which side, if intervention did become necessary, they would come down.

For Hungarians the Czechoslovak events were both good news and bad news. Having become a communist state, the republic could no longer make the persecution of ethnic minorities a national policy. While out of power, the Communist Party could support the process of ethnic cleansing as a tactical device, but it could not victimize Hungarian workers and peasants because of their nationality when it directed the affairs of the government. The bad news was that in the Cold War its geographic position determined a nation's political fate and no country east of the Elbe river was safe. What hope was there for Hungary to remain independent?

As the western powers saw it, Stalin's policy did not much differ from Hitler's totalitarian methods; hence it had to be resisted. Shortly after the Prague events representatives of Great Britain, France, Belgium, the Netherlands and Luxembourg met at Brussels and concluded a military alliance to oppose soviet expansionism. On June 1 the United States joined these European states in the so-called London Decisions that provided, among other things, for the convocation of a German constituent assembly and for the integration of the emerging western German state into the Marshall Plan. Inasmuch as the fate of Germany was the touchstone of soviet-western relations, the provision for a strong and viable Germany in the west escalated the Cold War to new heights. The Nazi menace that once held the Grand Alliance together was a thing of the past. The western powers had decided not to fight yesterday's battles but to concentrate on resisting the new enemy, the Soviet Union. The truism dating from

Neville Chamberlain's time that only a strong Germany could prevent a soviet march to the Atlantic was revived.

From the stiffening attitudes on both sides the soviet satellites were likely to suffer most. Hungary was early drawn into this new phase of the east-west confrontation. The Hungarian Communist Party (MKP) guardedly welcomed the intensification of the Cold War because it was likely to put an end to the ambiguity of Hungarian domestic and foreign policy. Mátyás Rákosi, on returning from a visit to Moscow at the height of the Czechoslovak crisis, had told his politburo that the press had to be instructed to emphasize that Hungary was now an active factor in the "peace front" (a recently coined euphemism for the soviet bloc). Until now it had been "but a card in the deck; now it had become a factor." Hungary's treaty with the USSR "filled the gap through which Anglo-Saxon imperialism hoped to move."[7]

But before Hungary could become a full-fledged people's democracy, the two workers' parties had to merge. In the gentler and more diverse political climate of the immediate postwar years the separation was tolerable and the programs of the two parties were recognizably different. But since then the USSR had become the sole point of reference for leftist parties everywhere and the division made less and less sense. Class-conscious workers—and opportunists as well—sensed the change in the political climate and responded to it with disconcerting eagerness. At the February 21 meeting of the MKP's political committee, Mihály Farkas, in the chair in Rákosi's absence, complained that in factories so many Social Democrats had switched to the MKP that in some places the Social Democratic Party (SDP) ceased to exist altogether. This was not what the communists intended. "If it continues," Farkas said, "the political balance will be lost." The SDP, he explained, far from remaining a workers' party, would join the opposition on the right. That would leave the MKP isolated and the noncommunist parties would array themselves against it. A hysterical fear of the dictatorship of the proletariat would erupt. Hence the goal was not the liquidation of the SDP but fusion.

A member of the committee moved that the party should declare a freeze on new members. Against this another member protested. So far, he said, "the dregs of the SDP had come over to us. Just when the better elements begin to join, it would be foolish to announce a freeze."[8]

His move against a freeze was carried. A little more than a month later the two parties fused and the problem of crossovers was solved.

After the abrupt sovietization of Czechoslovakia, Hungarian foreign policy entered a twilight phase that defies precise analysis. The ministry, although headed by a communist, Erik Molnár, remained moderate and showed no inclination for an outright eastern orientation. But internal factors militated against a nonaligned position. The one agency under firm communist control was the political police. More by happenstance than by design, the chief targets of the police soon became foreigners, usually businessmen taking advantage of the narrowing but still potentially lucrative opportunities the recovering Hungarian economy offered. Some of these entrepreneurs unquestionably employed less than ethical practices and often ran afoul of the law. The arrest of such a person had immediate political consequences that the foreign ministry was called upon to handle. Thus the control of foreign affairs began to slip from the hands of the ministry and became subject to coercive communist methods.

The first unhappy incident did not involve a businessman and there were no arrests—still, the repercussions were grievous. According to the American account of the incident, a train carrying Hungarian expatriates from Germany to their homeland was held up by Hungarian border guards at the Austrian frontier and the American personnel that accompanied the group was roughly treated. The American legation in Budapest protested; when it received no satisfaction, Washington halted further restitution of displaced assets from its occupation zone in Germany.

There were of course other areas of friction as well. The compensation of American individuals and enterprises for wartime losses had not yet even begun, while business firms in which Americans had a majority interest were being nationalized. It made for a deteriorating political atmosphere. Selden Chapin visited Erik Molnár on May 3 and found the latter sincerely interested in improving relations. Chapin reported home that it was "abundantly clear that Hungarian government greatly concerned over suspension of restitution shipments is searching desperately for some means of exit."[9]

Actually the government was more defiant than desperate and from this point on the tenor of diplomatic exchanges stiffened. Some communications from the western legations were left unanswered for weeks or months and in certain cases were not answered at all. American investment in Hungarian enterprises, so eagerly sought after in the past, was now discouraged. This, as Washington saw it, not only limited the scope of American enterprise, but also retarded reconstruction. Walter Bedell Smith in speaking to Molotov on May 5 cited this fact as one reason why his country had a growing interest in the defense of West Europe. He was particularly critical of attempts by "political minorities to seize power in certain states in the interest of an outside power." Such a development, Smith concluded, was bitterly disappointing to the American public—the door however still stood open for the settlement of differences.[10]

Molotov seized on that last phrase when he gave his answer on May 9. He declared himself gratified that the United States was ready to negotiate differences but rejected the charge that the differences stemmed from soviet interventions in East Europe. The treaties the Soviet Union had concluded with the people's democracies served the purpose of guarding against renewed German aggression and did not contain secret articles. The blame for the worsening international climate lay with the United States; here Molotov cited the Marshall Plan and developments in Greece as examples inimical to the USSR.[11]

Repeated references to Greece were unmistakable reminders of the spheres-of-influence arrangement, which, whether the United States accepted it or not, was an integral part of the new international system in Europe, a system the Soviet Union respected but the western powers repeatedly tried to breach. When for instance Washington, jointly with His Majesty's Government (HMG), demanded that Hungary give an account of how it was implementing the military clauses of the peace treaty, it obviously was not out of concern over the military threat Hungary posed, even if it overstepped the limitations of the treaty—which it did not. The foreign ministry in Budapest rejected the note with the argument that any information about treaty compliance had to be requested by all the powers represented on the Allied Control Commission (ACC), since the USSR did not join the British-American note, its demand could not be satisfied. The British were in any case uninterested in this diplomatic skirmish. The Foreign Office told the American chargé that "it would be unrealistic to believe satellites have any intention whatsoever of fulfilling obligations under treaties . . . if treaties were properly implemented it would mean dissolution of communist authority in those countries in which regimes are built solely on illegal and unconstitutional methods."[12]

The one concern that the British shared with the Americans was that satellite countries might overstep treaty limitations in order to augment soviet military might and the political control that went with it. They had information for instance that there were secret protocols in the soviet-Rumanian treaty that, when implemented, would have the result that "Rumanian army will be infiltrated by Russian officers, the cadres of Rumanian troops will go to Russia for training, and that the Rumanian army will be incorporated with [sic] Russian army to the limit of its forces."[13]

Self-interest however acted as a brake on the indignation of the western powers. They did not want to press their importunities so far that the soviets would feel entitled to similar information about Italy's state of military readiness; the resulting disclosures would have been embarrassing. In a garbled message to Chapin (as well as to envoys in Bucharest and Sofia, and repeated also to London) Acting Secretary of State Robert Lovett wrote: "We feel to establish now any precedent in Balkans involving recognition of right to inspect or obtain information unilaterally would substantially weaken our position in supporting Ital Govt in event Sov attack on its performance under military clauses."[14]

These exchanges on military and political matters still excited far less controversy though than questions of a cultural and religious nature. These latter were the items that in western minds served as the true test of the degree of soviet control. In Hungary they remained the center of public attention, as the religious question (with which the cultural was tightly bound up) was not one the authorities could deal with through regular means of enforcement. The Communist Party looked to Moscow for guidance and evidences were again contradictory. Moscow discouraged religion but did not suppress it and interfered with its practice only sporadically. The Hungarian legation in Moscow reported on a Greek Orthodox celebration of the Resurrection that drew crowds so enormous no church could hold them; the government, far from banning the celebration, allowed worshippers to spill into the streets. The writer of the report added that in Russia the Orthodox Church had always been a state church and now the government sought to use it to promote patriotism. "It might be good," the writer advised, "to propagate such freedom of religion in Hungary too in order to calm the unrest in religious circles."[15]

On the other hand the Marxist-Leninist view of religion remained the dominant factor in state-church relations in the USSR. Gyula Szekfü reminded the foreign ministry that, in Lenin's words, "in its relation to the socialist proletarian party, religion is not a private matter." The forces that comprised the Communist Party "cannot be indifferent to the lack of awareness, the obscurantism that religious beliefs signify." It was the duty of the party to point out the erroneous basis of religious faith, because the Communist Party was built "on a scientific, materialistic world view which unconditionally embraces the propaganda of atheism." Stalin firmly held that communism and religion were incompatible and that the party could not be neutral toward religion.

At the same time, in Lenin's view, there had to be a difference between the relation to religion of the state and of the party. He had written: "We demand that religion be a private matter in relation to the state, but under no condition can we regard religion as a private matter in relation to our party." In a state based on Marxism-Leninism such a distinction was of course difficult to put into practice. It could mean only that religious loyalty that did not violate loyalty to the state would not meet obstacles. Both Lenin and Stalin stressed that religion could not be fought by administrative (i.e. police) methods, only through education and a propagation of scientific principles. "The idea of 'otherworldly happiness' will no longer be necessary because with the cessation of man's exploitation by man and with . . . the development of technology, the achievement of a freer and better life on earth had become possible." All this didn't mean however that religion would not remain a problem. As Stalin pointed out, people's ideological development lagged behind their economic development and in socialist countries the influence of capitalist ethos would continue.[16]

This assessment of church-state relations was eminently applicable to Hungary. Religious practice did not meet obstacles as long as it did not meddle in politics—but it did, and with its fervid defense of the principles of private property it was the main exponent of the capitalist ethos. Rákosi's speeches acquired almost apostolic fervor when the subject was the incorrigibility of the church. In one of his speeches he threw down the gauntlet: "[W]e will not tolerate the abuses in denominational schools; we will oust reaction from the stronghold it is trying to build under the cloak of the Roman Catholic Church."[17]

Cardinal József Mindszenty was the main target of every such verbal attack. Had anticommunism not been at the core of his credo, his popular appeal would not have been nearly as great, for he was a pugnacious and meddlesome person with a rather transparent passion for making himself into a martyr. Of him the current envoy in Washington, Rusztem Vámbéry—certainly no orthodox Marxist himself—had said that he was a reactionary, opposed to agrarian reform, which had cost the Church a good portion of its property. His bishopric alone, by Vámbéry's account, had owned 87,000 acres. But he was greatly popular with adherents of the old order; his name was mentioned together with that of Miklós Horthy, and of Béla Imrédy, a former pro-Nazi premier.[18]

Western democracies did not look with favor on churchmen of this ilk, but in the Cold War they were potential allies. These men on their part looked on Americans in particular as a source of succor. Chapin was taken aback when a churchman appeared at his office unannounced and, after relaying Mindszenty's concern over the multiplying attacks on him, asked whether in Chapin's opinion the Cardinal should prepare for a short or a long-term accommodation with Communists since war was apparently inevitable.[19] Chapin avoided answering the primary question by taking issue with the secondary, that war was inevitable. Writing home, he asked for "such comments as the Department can give me for transmission to Cardinal."[20] Before a reply arrived Mindszenty made a second communication, asking for American intervention in the defense of church schools, which were being threatened with secularization. Chapin cited traditional American reluctance to intervene in such matters; Washington supported his position. "We do not of course underestimate importance of Catholic Church in current political scene in Hungary, but believe Vatican can more correctly assess situation on which to advise Cardinal with whom it apparently has adequate communication."[21]

The causes of this American reserve were deeper than tradition. Issues separating the eastern from the western camp were multiplying and the State Department was reluctant to add more. There was also the fact that in Hungary it was the church that repeatedly violated the principle of separation of church and state and did it in a combative manner that was difficult to condone and support. Cold War expediency went only so far; Mindszenty and his clergy tried to go a long step farther.

Chapter 19

The Break with Tito

It is no exaggeration to say that the unity of the soviet bloc, tenuous as it was in its early stages, was due almost entirely to Stalin's commanding authority, which brooked no resistance or disagreement. The cult of personality, for all its repulsive features, was paying dividends. Stalin towered over his contemporaries by projecting unbreakable strength and the ability to inspire terror, which, through the strange alchemy of emotions, was transformed into love in the hearts of millions. In his own sphere he acquired the status of a demigod and his (often illusory) achievements were ascribed to his preternatural foresight and uncanny genius. Not even Hitler at the height of his career elicited the extravagant praise and adulation that the recessive and laconic head of the soviet empire did.

When in June 1948, after years of probing, the Hungarian Communist Party (MKP) and the Social Democratic Party (SDP) finally merged into the Hungarian Workers' Party (MDP), "ending the decade-old separation of the Hungarian workers' movement," the leadership of the new party sent its first message to Stalin. Couched in the obsequious phrases of such documents, it conveyed "our conviction that under your wise leadership the peace policy of the Soviet Union will ensure the peace and free development of Hungary."[1]

It was ironic that at this time the wise and peace-loving Stalin, chagrined by a resurgence of German power as he saw it, as well as by dissonant tendencies in his own camp, ignited not one crisis but two, both of which in the end, far from enhancing his status, revealed his proneness to miscalculation and his inability to keep his power sphere monolithic after all.

After the breakup of the four-power Kommandatura that, according to the Potsdam agreements, was supposed to govern Germany, Berlin was no longer a capital city; yet it was here that the growing tensions attending the division of Germany converged. In West Germany the French recently had added their own zone to the already united British and American zone and the geographic outlines of a united Germany west of the Elbe were complete. The London Decisions had made this area eligible for Marshall aid; but the financial chaos produced by a plethora of currencies continued. The three western powers decided to introduce a new single monetary unit into Trizonia to put its commercial life on a viable basis. This posed a challenge to the USSR, which all along had desired a divided, impoverished, and politically disorganized Germany.

There was little Moscow could do to prevent the monetary reform in the western part of Germany, but when it was announced that the new currency would be introduced into West Berlin as well, Moscow rose to the challenge. Berlin lay deep inside the soviet zone of occupation and access to it had not been guaranteed in the four-power occupation statute. On June 24 soviet authorities, referring to technical difficulties, blocked traffic by land and water routes to the city.

Against this arbitrary act the western powers displayed an impressive unity of purpose that was itself embarrassing to Stalin, especially as in the first two years of occupation France had often sided with the soviets against Britain and the United States. France to be sure was still not reconciled to an economically strong Germany but in its interminable draining struggle in Indochina had become so dependent on American aid that it could not afford to antagonize Washington on the German question. American capital had scored another triumph. It was all the more essential that the soviet sphere demonstrate the same unanimity and firmness. Yet just now one nation in that sphere, the one that had been held up in Warsaw as a shining example of how in a people's democracy the Communist Party should absorb and liquidate its coalition partners, incurred Stalin's displeasure for repeated disobedience.

Some of the blame, to be sure, lay with Stalin, who was often imprecise in his desiderata. But as Stalin was infallible, Josip Broz Tito was declared guilty of breaching communist discipline. Stalin had for instance several times changed his mind over whether he should permit Yugoslav intervention in the Greek civil war, an act that would have been a violation of his sphere-of-influence arrangement with Churchill, or whether he should sanction Tito's leadership in the Balkan League that he had formed with Bulgaria and Albania to counter western penetration of the Balkans. But Tito had been lax in certain internal reforms Stalin deemed necessary if Yugoslavia was to become a soviet state, notably collectivization of the farms. To be sure, neither Poland nor Hungary had thus far undertaken this most difficult task, but in other respects their subservience to Moscow could not be questioned. Their communist parties waged a vigorous campaign against nationalistic tendencies; by contrast Tito's Yugoslavia was too Yugoslav to suit Stalin. His battle of wills with Tito in the end became a personal one; having silenced all his opponents within the Soviet Union, he was not disposed to put up with any in the satellites. And so at the end of June 1948 the Cominform met in special session in Bucharest and, under soviet pressure, expelled the Yugoslav party from its ranks.

The shock waves produced by this sudden break were too far-reaching to deal with in this study. Nevertheless repercussions in Hungary provide a microcosm that reflects all the features of the major struggle.

There was no advance warning, not even an intimation of why an emergency meeting of the Cominform had been called. On June 23 a front-page editorial in *Szabad Nép* had dealt with the West German currency reform and went to lengths to describe the dismal conditions in the western occupation zone that made the monetary reform necessary. Inflation was reported to be staggering, especially on the black market where most Germans bought their necessities. No German monetary instrument had any value, *Szabad Nép* asserted, only the dollar and gold, and, more commonly, cigarettes. General Lucius Clay, the commander of the U.S. occupation forces, had received suggestions that the new German currency should be backed by 50 million packs of American cigarettes.[2]

On June 29 the paper carried two articles hailing Hungarian-Yugoslav friendship. One article reported that a plaque had been dedicated the day before to the Yugoslav writer Yakov Ignatiev on the hundredth anniversary of Hungary's War of Independence against the Habsburgs. In this struggle, one speaker at the ceremonies reminded his audience, Ignatiev, "while preserving his loyalty to the Serb nation, has fought with Hungarian revolutionaries against Habsburg imperialism." Another speaker called attention to the "historic bonds between the Hungarian and Yugoslav nations and called the plaque "a symbol of our unshakable commitment to the preservation of our freedom and to its great guarantor, the Soviet Union, liberator of nations." The other article took note of the "important role the Yugoslav People's Democracy played on that section of the front of peaceloving nations on which it must stand guard."[3]

June 30 was the paper's day off. One July 1 a masthead editorial announced the resolution of the Cominform, condemning the Yugoslav Communist Party for following a wrong direction. "Many will ask," the editorial posed, "whether, before the split between the Cominform and the Yugoslav Communist Party became public, everything possible has been attempted to iron out differences in tested comradely fashion. The answer is, yes, we have tried everything to make the Yugoslav leaders see reason, to prevail on them to exercise self-criticism, to rectify their mistakes . . . to turn back from the fatal road they have taken. All in vain. Not only have they stubbornly kept their incorrect political line, not only did they defend or deny before [the rank and file] of their party and their nation the mistakes they have made, but they have gone against the great principles of the international working class and removed themselves from the . . . socialist front."[4]

The Central Committee of the Hungarian Communist Party met on June 30 for the express purpose of discussing the resolution of the Cominform. Its own resolution stressed the necessity of drawing serious conclusions from what had transpired in Bucharest because "in Hungary too there are, in the cities and villages, capitalist elements and the communist and social democratic parties have not yet fully fused. . . . A substantial portion of our party is on a low theoretical level and therefore we too face the danger of bourgeois nationalism, of being carried away by success, of distancing ourselves from the USSR . . . of the opportunistic illusion that a peaceful maturation to socialism is possible."[5]

The last phrase was grimly programmatic: it held up the Yugoslav defection as proof that socialism (let alone communism) could not be achieved without struggle and that peaceful evolution was an illusion. War had to be waged not only against bourgeois nationalism but also against such communists and Social Democrats whose intellectual understanding of Marxism was too primitive to make them effective in bringing the revolution to a successful conclusion.

One thing was beyond any doubt—that the Cominform had acted on Stalin's direct orders. The rationale was plain enough: Stalin was the guarantor of socialist unity, hence any challenge to his concepts and his will had to be quashed lest other leaders asserted their independence and ruinous diversity ensued. On the other hand, Karl Marx had held that while his principles were inflexible, their implementation had to take into account geographic, climatic, and historic factors. In tactical matters communist leaders were thus supposed to be free to chart their own course—but Stalin's authority negated that freedom. Stalinism by now had come to mean much more than merely his interpretation of Marxism; it also meant the unscrupulous application of terror against enemies within and without the party. His hold on communist parties abroad could be complete only if both his ideology *and* his methodology were applied.

Mátyás Rákosi could speak with authority on that subject. In his younger years, even though a fierce revolutionary, he had exhibited finer and gentler traits, had submerged his ego in the revolutionary struggle, and had showed genuine concern for the dispossessed. He suffered 18 years of imprisonment in dignity (and once went on a hunger strike in protest against the low wages his guards received). In 1940 he was freed and sent to the Soviet Union in exchange for some memorabilia from Hungary's War of Independence in 1849.

By the time he arrived in Moscow the great purges, in which so many foreign communists perished, had largely run their course. Rákosi grew into a leader in Stalin's shadow; he had daily contact with Comintern personnel that was itself Stalinist and looked on individual rights and freedom of expression as outdated concepts, incompatible with revolutionary necessities. He returned to Hungary in early February 1945 and unashamedly admitted that he had purged himself of all sentimentality. He was probably the most intelligent among professional Hungarian communists, but his mind was crammed with stock phrases about "reaction" and the social crimes of feudalism and capitalism, the corrosive influence of organized religion, and the inherent nobility of the laboring class. He frequently talked with Stalin on the telephone and presumably received direct instructions from him.

He had the opportunity to make himself popular (despite the fact that he was Jewish in a country that had a long tradition of anti-Semitism); his unprepossessing appearance, his simple habits, his humble roots, a certain crude charm, and a feel for the vernacular often endeared him to his audience. He was a Stalinist but did not emulate Stalin, never wore military attire, and did not cultivate Stalin's awe-inspiring remoteness. But his simplicity was a pretense; it concealed a hard, calculating party man.

Rákosi maintained close relations with foreign communists and had a special regard for Tito and the Yugoslav Communist Party. His private papers contain an article he apparently had never published on the occasion of Hungary and Yugoslavia concluding their cultural agreement. It was no accident, he wrote, that it was with this country that Hungary first established cultural ties. "Of all nations in the Danube basin it enjoys here the greatest sympathy. Part of it is our admiration for its heroic struggle for freedom [during World War II] and part our gratitude for the magnanimous policy of her great leader Tito, who has set past passions aside."[6]

In March 1948, only three months before the Cominform's break with Tito, Rákosi, Mihály Farkas and Ernö Gerö, the hard core of communist leadership, had journeyed to Belgrade for discussions on trade and other matters. (A hint that the discussions were not expected to be entirely trouble-free can be gleaned from the instructions the politburo of the MKP gave to the trio, that on all subjects relating to foreign and military matters they should be guided by the consideration that relations between the MKP and the Communist Party of the Soviet Union (CPSU) were of chief importance.)[7]

By this time complex questions of theory and practice had surfaced in Hungarian-Yugoslav relations. Tito, as we have mentioned in another connection, had introduced in Yugoslavia the concept of the "popular front from below," a euphemism for a communist-dominated grouping of progressive parties in which the presence of other parties was mere decoration; the Rumanian and Bulgarian parties adopted the practice, as well as the theory, and in the winter of 1947 gave the visiting Tito a rousing reception. In Hungary the political scene for such monopoly of power by one party was not ripe; apart from the vigor of the Smallholders' Party, Stalin still favored a coalition government. When Rákosi and his colleagues visited Belgrade, they probably intended to explain the Hungarian position.

Little is known of what transpired; most likely Tito, aware of what the Hungarian party did not even suspect—that his break was Stalin was imminent—was diplomatically noncommittal. The press made only routine references to the comradely spirit that had animated the discussions. What happened later, with a suddenness that was unusual even in the communist orbit, showed only that national parties (other than the Yugoslav) had no freedom of action and were not allowed to make adjustments for the political temper prevailing in their country; Moscow made a 180-degree turn and they blindly followed. From one day to the next Rákosi's respect and affection for the Yugoslav Communist Party became a thing of the past and he and his colleagues were immediately absorbed in the task of convincing the nation that the adulatory approval of the Yugoslav party had been a mistake, and one that had to be speedily corrected.

There was one aspect of this propaganda campaign that could not be delayed a minute. A corner of southern Hungary was home to a Titoist South Slav Union, organized along communist lines and separate from the MKP, with which it maintained friendly relations. On the morning of June 29, the day after Tito's expulsion, Farkas summoned the secretary general of the union, one Anton Rob, who was also a parliamentary deputy, and demanded to know whether Rob was in concord with the resolution of the

Cominform and whether he was willing to explain to members of his party the necessity of condemning Tito and the Yugoslav Communist Party. He handed Rob a batch of papers pertaining to the expulsion, allowing him one hour to study it and to declare his compliance with the request.

Rob at once refused to execute a volte-face like this. He explained that the Yugoslav Communist Party enjoyed enormous popularity among South Slavs in Hungary and until the day before had been held up to communist parties everywhere as the example to follow. He posed to Farkas whether members of the MKP, if asked to condemn Rákosi or Gerö, would do so.

Farkas was not inclined to polemicize. He demanded that Rob resign his parliamentary mandate and turn in his party membership booklet. The meeting ended in a stalemate and Farkas immediately initiated steps to expel Rob from parliament. When the news broke it produced consternation in the South Slav Union, which organized mass meetings to protest the high-handed conduct of the MKP. The Titoist rift was reenacted within Hungary; the central party dispatched a small army of agents to persuade the South Slavs to adhere to the Cominform's decision. A report dating from the end of July, issued by the head of this informational team, assured the leadership in Budapest: "Situation that has arisen in the South Slav Union in Hungary has, after three weeks of hard work, been reversed. We have taken the leadership from the hands of Anton Rob and his clique. The measures we have instituted assure that future leadership will remain firmly in the hands of our party."[8]

The speed and thoroughness with which "Titoism" was condemned and placed on the proscribed list was proof of the discipline Stalin had imposed in his sphere. The Hungarian envoy in Belgrade, Zoltán Szántó, who in the past had repeatedly praised Tito's concept of coalition politics, sent, one week after the break, a personal letter to Rákosi, with the obvious purpose of showing that he had dutifully shifted into ideological reverse. The Yugoslav party, he wrote, asserted that its condemnation by the Cominform was based on misleading information. Still, when facts spoke against the party, the comrades resorted to involuted arguments, totally at variance with Marxist-Leninist concepts. "They are fighting with their backs to the wall. They are bidding for mass support by fanning the flames of Yugoslav nationalism and chauvinism. Not only simple party members but functionaries as well stand in total bewilderment in the face of events. . . . They ask, 'What happened, how could we have come into conflict with the Soviet Union and the communist parties?' Old party members, mainly those who [during the war] had been in emigration, are in a virtual state of collapse; some look as if they had aged years."

The report spoke of arrests and suicides. The bourgeoisie by contrast had become less hostile to Tito. They reasoned that as the breach between the Yugoslav and the soviet parties widened, Tito would be forced to turn to the west and that would bring with it a reversal of nationalizations.[9]

The immediate reason for Stalin's break with Tito was the latter's deployment of troops in Albania for intervention in Greece, a move fraught with international complications. But the most prominent charge leveled against Tito was his slowness in collectivizing agriculture. Within a week of the Cominform decision the Hungarian government, itself lackadaisical in ending private ownership of land, introduced measures aimed against independent peasants. Taxes on them were sharply raised, the usual first step toward making private farms unprofitable.[10] All in all the Tito affair, while embarrassing to Stalin, speeded up the hitherto halting sovietization in several soviet-bloc countries and definitely marked the end of the temporizing tendencies Stalin had displayed thus far toward certain countries, most especially toward Hungary.

His instinctive reaction to every international development that he considered inimical to the USSR was to tighten control in his sphere, as, in his view, the United States was doing in its own. The campaign against Tito he could control; but the other crisis he precipitated in the summer of 1948 was a far more risky affair. The so-called Berlin blockade was an uncharacteristic move on the part of a man who usually displayed great discretion in international affairs. Even if the blockade succeeded the gain would be small: the liquidation of a western hedgehog position deep inside the soviet zone. But the Cold War was at a stage when even small victories counted for much.

The feeding and supplying of some two million persons in West Berlin seemed a hopeless task, so Stalin could be confident that the western powers would retreat rather than court war. But Harry Truman, who in tight situations could be as tough as Stalin and had at his disposal the vast technological reservoir of the most advanced nation in the world, showed no intention of backing down. While he rejected suggestions that the United States challenge the blockade by sending an armored train across it, he ordered an airlift, for the air routes were the only ones the soviets were unable to control. For many weeks the success of this bold enterprise hung in the balance as a large part of the population of Berlin came to depend on food and necessities flown in by transport planes hastily assembled from such distant parts as Hawaii, Panama, and Alaska. *Szabad Nép* carried daily reports declaring the American position in Berlin doomed, contending that chaos reigned in the city and in West Germany as well.

The Vatican also took an interest in the conflict, though for reasons unconnected with the larger issues. Among the articles taken from Hungary by the Germans in the last phase of the war was the Crown of St. Stephen. The first Christian king of Hungary had received the crown, with a cross on its helm, from Rome in recognition of bringing his barbaric nation into the fold. The crown was assumed to be in Berlin, at an undisclosed location. With the likelihood of the entire city falling to the soviets growing, the Vatican Apostolic Visitator in the United States addressed a letter to Robert Murphy, political adviser to the commander of American forces in Germany, requesting that the crown be transferred for safekeeping to the Vatican until such time as it could be returned to Hungary. A copy of the letter was forwarded to the American legation in Budapest. Selden Chapin opposed the request: "our legal right to transfer to the Vatican an object which belongs to the whole Hungarian people, on basis of claim it given 947 years ago by Vatican should be carefully examined," he wrote.[11]

It was and the crown remained in American custody for the time being.

These care-laden days were hardly propitious for convening an international conference to settle a question with a long and contentious history, that of Danubian navigation, but the peace treaties called for it and the revival of trade along inland waterways made it necessary. The Danube flowed across or formed the boundary of eight states from West Germany to the Black Sea. The conference had been called to Belgrade months earlier; by the time it convened the city was in the center of a bitter international dispute. But the matter at hand transcended doctrinal disagreements and the conference opened on July 31.

One of the interested states, Austria, lacked sovereignty and sent only an observer; Germany was not represented at all. The United States, Britain, and France had large delegations with full voting powers. The votes of all the other states were controlled by the Soviet Union and commanded an absolute majority. On the fourth day of the conference the soviet delegation moved that Danubian navigation should be the exclusive concern of the riparian states and they alone should sit on the Danubian Commission. The western powers spoke against the motion but were outvoted. Their motion that the decisions of the conference should be subject to ratification by the parliaments of the attending states was also defeated.

The final convention was signed only by states served by the river. It provided for the free navigation of these states and set up a supervisory commission composed of all interested states except Germany. To the surprise of all present the Yugoslav delegation voted with the soviet-bloc states on all major issues. In fact, when the French delegation, in refusing to be a party to the convention, called it a Diktat, the Yugoslavs indignantly protested. Other delegates took exception to being called members of a docile majority. In the end all seven attending riparian states voted for the treaty; the western powers refused to participate in the voting.[12]

The outcome represented a limited victory for the USSR at a time when it had little else to cheer about. The Berlin blockade became pointless as the airlift succeeded beyond all expectations and Stalin's feud with Tito had opened a crack in the Iron Curtain. This latter development, while encouraging to the west, also held dangers. The Policy Planning Staff in the State Department warned all missions abroad to be extremely circumspect in their dealings with Yugoslavia and not to lose sight of the fact that it was still a communist state "dedicated to an ideology of hostility and contempt toward the 'bourgeois capitalist world.'" It raised the point that American attitude "may have an important influence on whether the rift between Tito and Moscow spreads to Russian relations with other members of the satellite area or serves to weld those other members still more tightly to the Kremlin."[13]

The truly salient fact however was that Stalin's steely authority had been shaken. It is one of the dangers of a personal dictatorship that cohesion is provided by a single vulnerable bond. The satellite leaders understood that Stalin would go to any length to prevent that bond from further being weakened. In Hungary the process of sovietization accelerated at a faster rate than in other countries. The campaign against the kulaks intensified. Instances of friction with the western powers sharply increased. The Voice of America, which to many Hungarians provided an escape from the tendentious propagandistic news coverage at home, came under attack, leading to repeated exchanges between Washington and Budapest. Harassment of citizens and diplomats of Anglo-Saxon countries became commonplace. In these tussles of annoyance Hungary enjoyed all the petty advantages. Its own diplomatic missions, especially after the defections following the Nagy affair, were understaffed and the personnel too absorbed in the day-by-day conduct of affairs to have the time, or the sophistication, for subversive activities. The British and American legations in Budapest on the other hand had installed staffs of extravagant size and employed many Hungarians in service jobs. The disproportion was even greater in the numbers of businessmen and semiofficial functionaries on either side.

All this gave the Hungarians a kind of moral advantage, enabling the government to contrast the probity and exemplary conduct of its own representatives abroad to the shameful speculations and profiteering of western businessmen in Hungary, and their often successful attempts to enlist Hungarian entrepreneurs in their shady activities. The Budapest government knew very well that western countries were extremely sensitive about the treatment of their citizens abroad and when those got into trouble they were often ready to make major sacrifices to protect their rights and obtain their freedom. In the past Hungary's record on this score had been unexceptionable, but now communist officials unscrupulously sought to use this sensitivity to their advantage.

When the occasion arose to take legal action against a citizen of a western state, far from respecting his rights, Hungarian authorities tried to wring political, but mainly economic, gain from keeping him hostage. The consequences often hurt the Hungarian side more than the British or the American, as Hungary could ill afford economic sanctions against it, but by now the Two Camps concept took precedence over narrower considerations. The anger produced on either side by the sanctions and countersanctions was useful to the communists in widening the breach and legitimizing political repression. In the United States at this time communist-phobia ran deep, patriotism was often called into question, and left-leaning persons not infrequently had their careers ruined. In the communist camp suspicion went deeper, legal restraints were discarded, and the political police was given great latitude in pursuing its cases.

Small incidents escalated into interstate disputes. Each, while capable of resolution by local redress, caused progressively worsening relations. Early in 1948 there was another unhappy episode. A military attaché and an assistant military attaché of the American legation, while on a routine trip in western Hungary, were arrested by soviet military authorities, taken to Austria, and eventually, after intensive interrogation, released. The Americans at once protested, not only to the Hungarian government, but to the Soviet government as well. The point they raised was a somewhat pedantic one: did the soviets, present in Hungary no longer as occupiers but as supply troops, have the right to carry out police action against foreigners—and if they did not, as was obviously the case, should they or the Hungarians be responsible for the breach? In the exchange of notes that followed Moscow evaded the question with its usual obfuscatory tactics and nothing was resolved, but it was the last time the United States posed as a defender of Hungarian sovereignty against soviet encroachments.[14]

Before the eruption of the Tito affair such incidents could be smoothed over without damage to coalition politics—no longer. The Smallholders' prowestern tendencies now weighed heavily against them. Increasingly they were portrayed as agents of the capitalist-imperialist conspiracy. Western businessmen were characterized as parasites, and with some justification—their flashy lifestyle, their connections with Hungarians of dubious loyalties, and the foreign capital that stood behind them provided a glaring contrast to the general impoverishment of the Hungarian population. The press was intent on proving that these foreign entrepreneurs were agents of imperialist powers that tried to colonize every country lacking the economic means and political will to resist them.

All the dialectic skills of communist propaganda were enlisted in the effort. True to Stalinist practice, the security police (ÁVO) was the chief instrument of persuasion. It was, from top to bottom, a communist agency and the opposition had no means with which to fight back.

Chapter 20

Provocations

At the end of July 1948 President Zoltán Tildy was forced to resign. By decision of the Hungarian Workers' Party (MDP) politburo he gave as his reason an embarrassing occurrence in his "immediate circle" that was incompatible with democratic principles. It developed that Tildy's son-in-law, Viktor Csornoky, had become involved in questionable activities during his service at the Cairo legation. It was convenient in any case to effect a change at the highest (if largely ceremonial) level of government. Mátyás Rákosi charged the politburo with the task of choosing a successor.[1] As it happened, he had already made the choice: it was Árpád Szakasits, former head of the Social Democratic Party (SDP), which was now merged into the MDP. It will be remembered that in the immediate postwar period Szakasits had spoken condescendingly of the Communist Party, which in his view had developed strange and immature habits in illegality. Those habits might not have changed much since but they were the habits of the party in effective power. Szakasits's tenor changed from condescension to obeisance.

At the July 31 meeting of the politburo he announced that Rákosi had asked him to assume the presidency of the republic, a request that did him honor but did not make him extravagantly happy. Since it had been his honor (as a result of the fusion) "to set foot in this building, I realized what a party this party of ours is." He rhapsodized about his "liberation." Every day for him was a holiday, he said. "With what passionate love [our party] serves the cause of the nation, how high are its standards, what sense of responsibility it carries." And now he, a humble Social Democrat, had the privilege of sharing that responsibility. "I have never been so happy in my life," Szakasits exulted; "I can now be a part of this nation-building effort. This is why I am less than eager to be president of the Republic. I don't want to separate myself from the party, from the working class. I would be a thousand times happier to be president of the Hungarian Workers' Party than of the republic." But, he went on, as Rákosi had assured him that even as president he could continue to work for the party, he had consented to taking the position. Serving the party would inspire him, he concluded.[2]

In a roundabout way, Szakasits was speaking the truth. He realized he had been kicked upstairs and that serving the party meant placing his presidential powers at its disposal. What Rákosi had given he could also take away.

Meanwhile, even as members of the peace camp were drawing closer together, the difficulties with Czechoslovakia showed no signs of abating. Rákosi (to whom the problem was an extremely painful one) had daily reports of the worsening lot of Hungarians in the republic. His private correspondence contains a memo written by two Hungarian communists in Czechoslovakia, begging him not to forget that "the Czechs and the Slovaks in the past three years have followed a wholly nationalistic policy, considering only their own selfish interests, in defiance of the principles of the Communist Manifesto which [said that] 'communists . . . in the course of diverse phases of development . . . must always represent the interests of the common cause.'"

Czechoslovakia was not truly a people's democracy, the writers charged, and cited as proof the fact that it had been ready to participate in the Marshall Plan, that it refrained from making a friendship treaty with Poland because of the dispute over Tesin, and that it downright refused to enter into such a relationship with Bulgaria and set impossible conditions for a treaty with Yugoslavia. The Czechoslovak Communist Party, isolated from workers' parties in other countries, "drifted down the bourgeois-nationalistic slope and only after the February events [the communist coup] did it become aware of its rightist deviation. Only then did the Slovakian Communist Party change its thinking in the Hungarian question. It accepted with grinding teeth, under pressure . . . the agreement suggested by the Cominform."

In view of all this the writers held it to be of the greatest importance that the agreement drafted between Hungary and Czechoslovakia for the final resolution of the minority problem should be very precise, with no gaps that later could lead to new friction. This was all the more important because the problems with Yugoslavia made Hungarian-Czechoslovak cooperation imperative.[3]

Ten days later representatives of the Hungarian and the Czechoslovak parties met in Prague. The atmosphere was superficially felicitous. Deputy Premier Jan Siroky, who spoke for the Czechoslovak side, declared that the February events made an understanding possible. The Czechoslovak Communist Party wished to contribute to the solution of the minority problem. Beginning in September, he announced, Hungarians would be free to use their mother tongue, even in contacts with local authorities, would have the right to vote, and would receive back their confiscated land if it did not exceed 50 hectares. The law of labor mobilization would apply to all ethnic groups equally and Hungarians who had been removed from their places of domicile would be allowed to return. Hungarian-language schools would be set up within the Czechoslovak school system, with their own buildings, teachers, and supplies. Citizenship would be restored to those who had lost it and exceptions would be kept to a minimum.[4]

That these promises were not seriously meant became obvious soon enough. A few days after the Hungarian delegation returned home, Siroky spoke at a meeting of party secretaries and every pretense of goodwill was gone. Regarding the Hungarian question, he declared that the policy pursued so far had been correct and although it would be "in our own interest" that changes be made, such changes would have to have definite limitations. Hungarian schools would function as sections of Slovak schools; Hungarian press and cultural institutions would be under Communist Party supervision. Furthermore, re-Slovakization was not at variance with Marxist principles and would continue; far from victimizing Hungarians, it would return those among them who were once Slovaks to their roots.

Siroky admitted that ethnic purification was no longer possible but stated it could be partially salvaged through the population exchange. One participant suggested a solution under the heading of class struggle: Hungarians should be classified as reactionaries.

Rákosi learned of this meeting from the same two party members who had warned him to be careful about the wording of the agreement; the second message pointed out how right they had been in their pessimism.[5]

Negotiations between the two communist parties nevertheless continued well into August. Most of the discussion centered on the one problem that seemed to remain when all the others had been cleared away: what should happen to the Hungarians remaining in Slovakia after the population exchange. The Slovaks tried every means to prevail on Hungary to take them. They succeeded in enlarging the number of war criminals Hungary was to accept from 1,000 to 1,500; with family members the total came to 3,500. They also persuaded their Hungarian comrades to settle another 10,000 above those to be exchanged, on condition that the men were not older than 65 and the women were not older than 60, and both were fit enough to be transported and to work.

The pressure on the Magyars in Slovakia, after the spell of relief following the February events, was soon resumed. Hundreds were arrested and punished for speaking Hungarian in public places. Re-Slovakization was so intense that the rights vouchsafed to Hungarians were becoming meaningless as most of them, especially the younger ones, declared themselves Slovaks. The promise that those who had been transported to the Czech provinces for public work would be returned was not being carried out; many of those unfortunates wondered whether the mother country had totally abandoned them. The few who did return were not given permits to work in Slovakia. The "parallel schools" promised to be set up side by side with the Slovak ones languished for a lack of educators. The Communist Party, in admitting members from the Magyar areas, limited those to the re-Slovakized elements.

In fact, by every account, intolerance was most fierce among the communists. Gustav Husak, one of the leading Slovak communists, Vladimir Clementis, and a man named Okali, formed a clique dedicated to freeing Slovakia not only of Hungarians but of Jews as well. Confiscations continued and a report had it that the Slovak Commissariat of the Interior (in charge of the political police) had instructed lower administrative officials to prepare lists of Hungarians most likely to emerge as the cultural elite— they would be slated for deportation first (just as János Gyöngyösi had feared). Anti-Semitism ran deep, in the party as well as among the general population. In Bratislava a big anti-Jewish rally adopted the slogan: "Slovakia to the Slovaks, Palestine to the Jews." A number of former fascists were appointed to high positions in the Communist Party.[6]

The hatreds evoked by the war were slow to die. Remarkably though, in Hungary's own minority problem such negative emotion did not seem to be a factor. There was very little overt hostility toward the Swabians and the resettlement proceeded far behind schedule. The law that took such a long time to frame applied to 536,000 persons; of these 166,000 had been expelled by the end of June 1948. Then the process was halted. After deductions on a variety of grounds there still were some 230,000 Germans in Hungary who fell under the provisions of the law. Of these however only 35,000 were marked for deportation, while the others were in a state of prolonged limbo. Confiscation of land ran ahead of deportations and some

60,000 persons were left without any means of livelihood. The director of resettlement worried that these would become enemies of democracy. Many roamed the land and stole agricultural property, often from the very people who had received their farms. Because of the acute shortage of shelter many continued to live in the houses that were once theirs, sharing quarters with the new owners in conditions that led to endless quarrels. And this was just the time when the Office of Resettlement had to make provisions for accommodating hundreds of thousands arriving from Slovakia.

The director of the office seriously raised the question of whether the resettlement should continue at all. The only argument in favor of it was that otherwise many of the kulaks would retain their property while thousands of others, even small peasants, had lost theirs. Legal considerations were however overwhelmed by what was essentially a problem of poor relief. Dispossessed Swabians still in the country had to be provided for. Several solutions were being proposed, but each had more drawbacks than advantages. The indigent farm workers could be made tenant farmers. Or they could be employed as farmhands on properties distributed under the land reform. They could also be settled on farms vacated by departed Slovakians.[7] But the general impoverishment of the land frustrated all these schemes. And industrialization was too slow to relieve the pressure on tillable land.

Amid all these hardships, the general economic recovery was nevertheless impressive. By the middle of 1948 most of the scarcities had disappeared; the new currency, the *forint*, proved durable, inflation was negligible, and even the continued absence of the assets taken to Germany did not prove as catastrophic as had first been predicted. By suspending their return because of a minor insult the Americans deprived themselves of a potential means of pressure, which for Hungary was a negative kind of gain.

In September another incident imposed further strain on Hungarian-American relations. Although superficially a police matter, it involved the larger question of the autonomy of the Hungarian-American Oil Company (MAORT), the Standard Oil subsidiary that supplied nearly the entire petroleum requirement of the country. On the night of September 18 the company's president and a technical adviser were arrested and charged with economic sabotage. Selden Chapin immediately lodged a protest and the State Department upheld him. Acting Secretary of State Robert Lovett cabled Chapin that the detention of the two businessmen was a serious affair, "not only in itself but also as a precedent." The Hungarian government must be warned that "continuance of proceedings against [the two Americans] will entail material consequences."[8]

The two "saboteurs" were, after an exchange of notes, released and allowed to leave the country, but the American attitude toward Hungary hardened noticeably. And now the overarching question of how far Hungary should jeopardize its economic interests as it attempted to make a clear demonstration of its choice between the two camps emerged with compelling gravity. It is certain that at least two members of the central leadership, Rákosi and Ernö Gerö, were willing to go to any length in that direction and they were, as a matter of course, supported by Mihály Farkas. But the new foreign minister, László Rajk, who had given up the more prestigious interior portfolio to direct foreign affairs, possibly held different views. On November 4 he invited Chapin to his office and expressed himself concerned about the apparent American intention to "terminate all action on restitution from Germany to Hungary." He inquired whether relations could not be improved. Chapin replied that "essential first move would be withdrawal by Hungarian Govt. of its note of October 31" that called the suspension of the return of assets a violation of the peace treaty and threatened to demand the recall of those legation officials processing war damage claims.

Rajk reluctantly admitted that the note had been a mistake. He proposed conversations, first on the technical level, to find a way out of the impasse. Chapin offered the perfectly legitimate answer that he had to obtain authorization for holding such talks. To the State Department he wrote: "Although I am pessimistic as to possibilities obtaining any real satisfaction of our outstanding problems . . . I did not feel under circumstances that Dept. would wish to have me show a completely unyielding attitude." He suggested that if the department agreed to listen to the Hungarian proposals, it should make clear at the outset that those proposals would have to be concrete.[9]

This was just the problem though, that there was not one issue on which Rajk had anything concrete to offer. Nationalization of firms wholly or partly owned by Americans was well in progress and could no longer be halted, let alone reversed. The press campaign against "dollar imperialism" was in full swing. Western requests for information about the size and deployment of Hungarian armed forces had gone unheeded. At the Danubian conference Hungary had voted for the exclusion of the western powers from any right of oversight; this amounted to a declaration of solidarity with the soviet bloc. The list went on. What prospect was there for finding common ground?

Relations with the western powers had reached a stalemate and it represented a defeat of the policy the Smallholders had pursued since the birth of the "new Hungary." Though possibly it was a surrender rather than defeat. A good part of the Hungarian Workers' Party's (MDP) political clout was still more perceived than real. It had no substantial support from Moscow. It had achieved near monopoly of power not by being politically preponderant but by gaining control of the security police—and even that could have been prevented had the opposition closed ranks. Instead the opposition allowed itself to be fragmented and pinned its hopes on western support, for which the opportunity had passed. After the Cominform resolution the west concentrated on extricating Yugoslavia from soviet control; Hungary's rescue was in effect abandoned.

Chapter 21

Church and State

Within four months of the expulsion of the Yugoslav Communist Party from the Cominform the last vestiges of comradeliness between it and the workers' parties in other people's democracies had vanished. In the immediate aftermath there had still been grudging acknowledgments of the Yugoslav people's heroic struggle against fascist forces and of Tito's efficient liquidation of internal reaction. A towering figure like Josip Broz Tito could not from one day to the next be reduced to the rank of a disgraced heretic as Leon Trotsky, Lev Kamenev, and Nikolai Bukharin had been in the Stalinist purges. Zoltán Szántó, in one of his reports from Belgrade, conceded that Yugoslavia had been the people's democracy with the greatest prestige and "to gain her friendship and trust was a major foreign policy task." He commented with seeming approval on Belgrade's claim that the Cominform resolution amounted to a denial of the country's fight against the German occupiers and that it "defames tens of thousands of fallen soldiers and flings mud into the heroic face of the nation."[1] In another report Szántó noted that some communist parties were in no hurry to adopt the new line; the Central Committee of the Bulgarian party even stated that "the debate" could not influence the friendly relations between the two countries.[2]

By late fall all ambivalence was gone. Yugoslav diplomats in soviet-bloc capitals were being treated with studied discourtesy and their communications were given short shrift. When the Minister in Budapest, Mrazovich, about to be recalled, appeared at the foreign ministry, László Rajk, expecting an obligatory farewell visit, received him. But Mrazovich had come to protest against Hungary's violation of its reparations obligation and indicated that his government intended to submit the matter to an international forum. Rajk replied that if he had known the Minister came with a different purpose than to bid farewell he would have asked him to submit his complaint in writing. As for the charges, it was Yugoslavia who had violated the peace treaty and was now seeking to legitimize its violations through "the US-led imperialist voting machine [in the UN]." Rajk cut short Mrazovich's reply by rising from his seat to indicate that the audience was over. It had lasted only six to seven minutes.[3]

Mrazovich then sought out Rajk's predecessor, Erik Molnár, during whose tenure he had been accredited Minister. He came, he explained, to assure Molnár that there was not the slightest chance that the Yugoslav party would "slide over" into the imperialist camp. His party remained faithful to the Marxist-Leninist line and he was confident that in time things would return to normal. Molnár replied acidly that self-criticism was a better cure than time. He reminded his visitor that party discipline bound everybody, not only within a nation but in the international setting as well and that loyalty to the Soviet Union could not be the subject of an argument.[4]

But condemning Tito and "his clique" was not enough; after all the controversy could be portrayed as an honest disagreement among communists and a good number of comrades might even choose Tito's side, especially as his policies had been on the whole successful. The spectre of a separate Titoist faction haunted the precincts of international communism. It had to be shown, *proven*, that Tito had *never* been a true communist, that he had betrayed the cause from the start of his career, and that his efforts to sovietize Yugoslavia had been a camouflage to conceal his treachery. He had to be "unmasked," lest in the minds of some honest communists he became a victim rather than an apostate.

The campaign against "Titoism" became a major industry in every people's republic. The archives contain batches and batches of documents, "revelations" of such patent falsehood, that even the most devoted Stalinist must have had trouble believing them. Technically though the documentation was nearly flawless and it purports to show that Tito and his fellow-traitors (Alexandar Rankovich, Edvard Kardelj, Milovan Djilas and others) had been double agents who during the war had cooperated with the Germans; and that Tito had been an ineffective war leader, concentrating on military matters in which he was inexperienced, while political work was neglected. He had sought cooperation not with the Soviet Union but with the western powers. "Such strange conduct," notes one document, "seemed [in its own time] to be a puzzle. . . . Today we have the key with which [the puzzle] can be solved. In one word: treason. An old and sly traitor [ranged himself] against his nation and the socialist cause [directing] a whole little army of spies and imperialist agents."

Another paragraph charges that "Tito conducted his odd and criminal 'delaying war' by assigning to proletarian units pointless tasks; he moved and guided his troops according to instructions received from Anglo-Saxon spy organizations within the framework of the delaying war ordered by the western allies, whose primary goal was . . . to weaken the main troop formations, to prolong the war as long as possible, to prevent the sweeping victory of the USSR and at the same time preserve for the postwar period an Italy and Germany which, although defeated, [could participate] in a new military adventure against the USSR in the service of Anglo-Saxon imperialists."

Tito was also guilty of using the war to destroy the greatest possible number of true communists and patriots who were the most active champions of Yugoslav-Soviet cooperation. He had been active in preparing a second front in the Balkans, the one that Churchill had planned, but that never came to fruition.[5]

Graver accusations, that Tito had organized spy networks in the Soviet Union and the people's democracies, followed in due course; they led to a rash of spurious prosecutions in which hundreds, then thousands, of communists in the soviet bloc were caught up. This was in line with the Stalinist premise that the treason of one person could be eradicated only if the entire camp of its adherents, admirers, and accessories was destroyed as well.

The term "reactionary" now came to embrace turncoat communists, renegades who placed love of country above the cause of the international workers' movement or who questioned the liberating mission of the Soviet Union. The assumption was that these men had been enlisted and corrupted by agents of the reaction organized in an international network of subversion.

This ideological premise justified an increased struggle against church and religion. In the USSR the Communist Party had accomplished the formidable feat of dechristianizing "Holy Russia" by making atheism the official "religion"—if there still were superstitious souls in the old generation, those reared on the Marxist ethos were enlightened and contemptuous of tradition. The satellite countries were now expected to wage their own Kulturkampf.

This contest between political and spiritual forces was bound to be an unequal one: the churches drew their strength from enshrined tradition, while communism was essentially rootless and fatally experimental, still groping for a state system that could be made viable without coercion. Even in the Soviet Union the full extent of the party's success could not be gauged; terror was so pervasive that it was impossible to tell fear and conviction apart. In the satellite countries the nature of the struggle often depended on the personalities involved and on the pragmatism, or lack of it, of the leadership on either side. In Czechoslovakia for instance there scarcely was a struggle at all. The Communist Party leadership, in a secret report prepared in its inner circles, admitted that it could not at once "keep in touch" with the Church (as conditions demanded that it should) and carry on the fight against it. To build socialism, the party needed support from organizations with a popular base; it found it expedient to cooperate with the Christian People's Party, which did have such a base, being in charge of a number of charities and places of worship, and thus served as a link between the Church and the government. Members of this religious party, the report frankly admitted, stood "on a foundation of Christian principles [and] they build, under the leadership of the Communist Party, the road to socialism and, above all, they stand behind the struggle of the Soviet Union for peace, which is humankind's sole escape from destruction."[6]

Such a pragmatic attitude on the part of communists and churchmen alike was the exception. In Hungary by contrast the communists made no attempt to "keep in touch" with the Church and to use its organizations for their own purposes. Possibly, had the primate of the Catholic Church acted with greater discretion, had he not posed as the conscience of the entire nation, an open break could have been avoided. When the French ambassador to the Holy See had an interview with Cardinal Montini and sought to draw him by saying how he admired Mindszenty's firmness of character, Montini replied that the Holy See wished the Hungarian primate had been less pugnacious, especially on the question of schools. He further expressed the opinion that the Hungarian government would hesitate to arrest Mindszenty for fear of making a martyr of him and causing unrest in the country. The Vatican could never disown the cardinal, Montini concluded, but if he had asked for advice, he would have been admonished to be more circumspect.[7]

It was not only Mindszenty though; the Christian clergy as a whole in Hungary made no effort, as it did in Czechoslovakia, to find a modus vivendi with the state or to moderate its rhetoric; in church schools it continued to denounce communist atheism. The Catholic Church in particular cast itself in the role of a victim; its claim that it was being targeted for destruction proved a self-fulfilling prophecy. Its attempts to involve the western powers in the struggle, and its suggestion that a failure on the part of these powers to take a position was a shirking of moral responsibility, caused no end of embarrassment, especially to the English and the Americans.

His Majesty's Government took a rather uncharacteristic interest in the fate of the religious establishment in the people's democracies. It was frankly concerned that the communists felt strong enough to challenge the Church and even to engage in persecutions. Newspapers made a point of it that in Czechoslovakia the communist coup was carried out without opposition from the Church—they wondered whether that was a sign of surrender. In Hungary, as Fleet Street saw it, the struggle had not even begun and already the church had lost. The Hungarian legation in London, in reports home, commented on these observations and on one occasion wryly noted that the Church in Hungary had indeed lost its battle with the government, because "a very stupid man" conducted its affairs. Among other errors Mindszenty committed the fatal one of identifying the Church with the cause of Habsburg restoration, a policy the Labour government could never support. But why, the author of the memo wondered, did that same Labour government extend so much courtesy to Hungarian emigrés, most of whom were monarchist in their sympathies? It was politics, all the more blatant because support to an activist Church was at variance with the anticlerical tradition of the British government.[8]

The clergy within Hungary meanwhile sought support for its cause mainly from the Americans. It will be remembered that in April Mindszenty had asked Selden Chapin's advice as to how far he should go in accommodating the communists; he had found the envoy unresponsive. In November an eminent prelate visited Chapin and used a different approach. He said the Catholic masses in the country had repudiated the west and were showing preference for an eastern orientation.[9] The visitor had a plan for arresting this trend. Chapin should send an emissary to Rome to persuade the Holy See to appoint this visiting prelate as special envoy to Hungarians living in the west (presumably to counter the blandishments of the Hungarian government).

Surprisingly, Chapin supported the request, arguing that if the person he planned to send to Rome went in a private capacity, it could not be construed as government intervention; he had in fact already authorized the trip on the ground that "the consolidation of Church resistance to Communism is in our general interests."[10] (The department differed, writing in reply that, "aside from inappropriateness US interference Church affairs . . . premise disunity Hungarians abroad not confirmed by facts situation."[11])

On November 17 Chapin visited Mindszenty at the latter's request and later reported that this had probably been his last conversation with the cardinal, who expected "action of some sort against him" daily.[12] Mindszenty pleaded for American intervention, not on his own behalf but on behalf of the priests and members of religious orders who were reported to be under the threat of internment and deportation. Chapin somewhat lamely suggested that the cardinal address his request to the Vatican.[13]

As the year wound down events accelerated as neither the Church nor the government showed any inclination to back down. The latter circulated a petition inviting people to condemn Mindszenty for antistate activities. On November 21 the cardinal urged believers, in a message broadcast from all Catholic pulpits, to sign the government's petition, because refusal to do so might cost them their jobs and livelihood.[14] At the same time he sent a letter to a newspaper of a moderate stance, vowing that threats by the government would not deter him from fulfilling his duty, adding that he had resigned himself to whatever fate was in store for him. His pledge was published, but the government confiscated the issue of the paper in which it appeared.[15]

Mátyás Rákosi did not shrink from the confrontation. Addressing the central executive committee of his party, he devoted much of his speech to church-state relations. "The people now fault us," he said, "for showing such weakness in the face of Mindszenty's fascist religious services. Workers complain about the abuse of power by the clergy toward their children. We are willing to meet every attempt to find a solution. If however an agreement is not possible, we will defend democracy—as the people demand that we should—and we will strike not only against the lower clergy but against the responsible higher clergy as well."[16]

The time for reconciliation had passed. On December 23 Mindszenty was arrested, together with 13 suspected coconspirators, on charges of treason, spying, and illegal currency trading. The arrests were made public only on December 27, after the Christmas holidays.

To international opinion it barely mattered whether the charges had any foundation in fact; the issue was not Mindszenty's guilt or innocence but whether a state's authority could be used to crush a religious establishment. The Budapest government was ready to confront this question head-on. It was well aware that foreign disapprobation of a government's internal action usually lasted a short time, until it gave way to practical concerns; since in the case of the cardinal a weighty principle was at stake—could religious freedom shield individuals and organizations from the charge of subversive activities?—the price was well worth paying. Chapin offered it as his opinion that the government would not have made the arrest "except on orders from Moscow." He added: "Kremlin immediate program includes early dissolution religious and other remaining ties of eastern with western Europe."[17] His advice was that "Mindszenty case should not be treated as isolated example violation human rights. But linked to all forms of persecution, so that his arrest becomes symbol of destruction human liberties. Case Mindszenty particularly felicitous for such exploitation in eastern Europe, since average man of whatever denomination whose abstract thinking largely limited to symbolizations in terms concrete personalities and events, has already come to regard Cardinal as prime symbol of Western ideas of liberty and Christian values."[18]

Acting Secretary of State Robert Lovett agreed with this evaluation but thought there was "minimum likelihood effective intervention his behalf."[19] The view was gaining currency in the west that official interventions would do more harm than good. Past experience showed that the sentences meted out after such meddling from abroad were generally harsher than they otherwise would have been.

It has been argued that the Budapest government staged these judicial proceedings to divert attention from the internal problems created by sovietization. However this is a simplistic theory that ignores the crucial role of priorities in the communist system. Problems of an economic nature could be adequately dealt with only if the ideological groundwork was in place. It was not enough to partition the large estates, to nationalize business enterprises, to fix the price of staples—other economies had done that without going all the way to socialism. It had to be shown that these achievements gained their legitimacy by the destruction of the old order and the consequent emancipation of the working class. The silencing of a reactionary Church was necessary for the fulfillment of the multiyear economic plans with their astounding statistics.

That Hungary simultaneously was able to destroy the old and construct the new undoubtedly impressed Stalin. Together with the increased tempo of agricultural collectivization, this ability provided an exemplary contrast to the Yugoslav heresy and the slow-paced recovery in other satellite states. If Rákosi's version was correct, the final sovietization of Hungary had to await the economic collapse in the west. But that event was an ever-receding prospect and Hungary could not be kept in an ideological limbo indefinitely. Evidence suggests that the Marshall Plan had been a rude eye-opener for Stalin. Not only did it reduce the chronic American production surplus but it also created new markets and reaped rich political benefits. Now the Berlin airlift, an unbargained-for consequence of Stalin's rashly instituted blockade, produced its own miniature Marshall Plan with thousands of tons of goods being flown into that city daily. Had the blockade come in winter, this impromptu relief by air could not have met the need for food, fuel, and clothing; the warm weather kept demand low and the "chaos" over which the communist press gloated almost daily never materialized.

Such an exhibition of economic and logistical efficiency was naturally unsettling for Stalin. He deemed it necessary to produce a Marshall Plan of his own. On January 20, 1949, in Moscow, the Council of Mutual Economic Aid (COMECON) was formed with the USSR, Poland, Czechoslovakia, Hungary, Rumania, and Bulgaria member states.

The pact was actually the culmination of a long-maturing process. According to the U. S. chargé in Moscow, leaders of the bloc nations had, at soviet behest, decided at a meeting in Sochi in the Caucasus the past November to integrate their economies and inaugurate "some form of ruble area with satellite monetary and banking systems brought into more direct relationship and dependence operations State Bank USSR." In a later report the chargé took a somewhat exaggerated view of the intended function of COMECON: "to remake [satellite countries] . . . on exact pattern Soviet republics already within Soviet Union . . . until they are for all practical purposes active part of Soviet Union directed from Moscow."[20]

By the rules of COMECON, membership was open to all applicants; Yugoslavia was the first country outside the founding group to apply. In a note to the foreign ministry in Budapest it solicited Hungary's support, referring to the friendship treaty of the year before. The request was coldly rejected on the ground that Belgrade had not reduced its reparations claims on Hungary.[21] The reply further charged that the Yugoslav government had confiscated Hungarian goods on its territory and had persecuted ethnic Hungarians sympathetic to the Soviet Union; it ended by saying that until such time as Yugoslavia had mended its ways, it had no prospect of being admitted to COMECON.[22] Other satellite governments sent similar answers, enabling the Tito government to charge that the parallel rejections were proof that its bad relations with the people's democracies were not Yugoslavia's fault.

In Hungary meanwhile the Mindszenty case stood in the center of all politics, publicity, and speculation. The French envoy told Chapin that according to his information the case had been prepared in and directed from Moscow, which wanted to force a showdown on the religious question in East Europe.[23] The counts of indictment against the 14 defendants were so extensive that they drew people of the most diverse backgrounds and stations into the network of complicities. The government issued a Yellow Book, in English, French, and Hungarian; some of its documents even implicated Chapin. He was accused of trying to get Mindszenty to spy for the United States to obtain information about the soviet occupying forces, and of conspiring with him for the restoration of the Habsburg monarchy. The cardinal and his accomplices were reported to have admitted all this under questioning. The government now had the attractive choice of declaring Chapin persona non grata and forcing his recall, or letting him continue at his post exposed, discredited, and cut off from his contacts.

Reluctant to provoke a confrontation, the government chose a milder course. It handed Chapin a note (dated January 29, 1949) demanding that an official of the legation, Secretary Stephen Koczak, suspected of carrying out intelligence work, leave the country within 48 hours. The note emphatically stated that the expulsion was not connected with the Mindszenty case.[24]

Charges of this nature against diplomatic personnel were becoming commonplace, not only in Hungary but in other satellite states as well. The apparent intent was to create the impression that the western powers were trying to infiltrate and subvert the body politic in the people's democracies and that a good part of the west's diplomatic establishment was engaged in espionage.

The view in Washington was that the United States should, if only for prestige reasons, take retaliatory action. John Hickerson, director of the European Affairs Office, consulted the legal department regarding the possible expulsion of the first secretary of the Hungarian legation, János Flórián. Hickerson pointed out that Flórián had intimidated several former members of the legation who had defected and was suspected of being a representative of the Hungarian "secret police." However, Hickerson recommended that no connection should be made between the Flórián and the Koczak affair; "we should avoid official adherence to reciprocity principle which the Hungarians could extend to our disadvantage."[25]

Expulsions from the Washington legation were obviously a problem. Hickerson's deputy, Llewellyn Thomson, noted that "the Hungarians have only three [accredited diplomats] and . . . we may run out of new material shortly when they will be down to one and our alternative will be to submit to their picking off our people one by one or breaking relations with them."[26]

Mindszenty's trial, between February 3 and 5, created a worldwide uproar. Endre Sik, the new Hungarian Minister in Washington who had replaced Rusztem Vámbéry, later wrote in his memoirs that from the opening hour of the trial the legation did not have a single peaceful moment. American newspapers hailed Mindszenty as a martyr, and angry crowds kept the legation under a virtual siege for hours. On the last day of the trial Catholics in New York held a huge rally on Fifth Avenue, with some 100,000 people demanding the cardinal's release.

Deliberately or not, the trial was held in a rather small courtroom and many requests for tickets were denied. The British legation repeatedly complained for not being given the opportunity to observe the trial; it called the whole affair a violation of the peace treaty. The foreign ministry replied that the treaty did not give the powers the right to interfere in Hungary's internal affairs and that, furthermore, it did not oblige Hungary "to safeguard fascist and anti-democratic elements, but on the contrary, to eliminate them."[27] On February 8 a communist publicist of some renown, in *Szabad Nép*, reacted indignantly to Ernest Bevin's charge that Hungary was violating human rights. "We Hungarians," he wrote, ". . . find it repellent that the Labour Government, which everywhere from Greece to Malaya oppresses human freedom, should presume to criticize a nation and a judicial process as to whose fairness there cannot be any doubt."[28]

The People's Court sentenced Mindszenty to life imprisonment; his alleged accomplices received prison sentences of varying lengths. There were immediately calls for the U.S. government, President Truman himself, to intercede on behalf of the convicted. The newly appointed Secretary of State, Dean Acheson, called for extreme circumspection lest the cardinal's case be further harmed. President Truman, speaking to the press, agreed with this assessment and referred to Hungary as a police state and to the tribunal that convicted Mindszenty as a "kangaroo court." He added that the Hungarian people were not responsible for their government's action and that the United States government was investigating whether the trial had been a violation of the peace treaty.[29]

It was diplomatic warfare in its least civil version, each side seeking to garner maximum advantage from whatever fact or circumstance served its case best. Endre Sik called on Dean Rusk, assistant secretary of state for United Nations (UN) affairs, and inquired what the United States government proposed to do about Chapin's involvement in the Mindszenty case. But there was of course no proof of the Minister's involvement and Rusk pointed out that Chapin's request for a transcript of the trial was never honored.[30]

On February 11 Hickerson recommended that the department reject all accusations against Chapin and declare its complete confidence in him. But on that day Endre Sik delivered a note stating that his government had declared Chapin persona non grata and requested his recall. In its reply the department informed Sik that Chapin would be called home for consultation but that no decision would for now be made about his official position.[31] (Chapin left Budapest on February 17 but it was not until May that he tendered his resignation, which President Truman accepted. Four months later Chapin was named ambassador to the Netherlands.[32])

The day after he left Budapest, Chapin said at a news conference in Paris that "no one today, except the blind and the twisted, can fail to see that the Hungarian people is under the complete, total domination of a group of Moscow-trained Communists whose sole allegiance is to the Kremlin."[33]

Against this and similar accusations Moscow argued that soviet control, to the extent that it was exercised, meant not domination but the liberation of the laboring classes from the feudal-capitalistic system that for centuries had exploited them and that the United States for its part dominated the states of West Europe for selfish purposes. It should be noted that at this early stage of sovietization there was some validity to the first argument. The land reform and the outward forms of dignity bestowed on the working man had improved his condition and elevated his self-respect. That the process was accompanied by the gradual silencing of all opposition, and by judicial measures of doubtful propriety, could be defended on the ground that the process of transition demanded disarming the forces of the old order lest they attempted a comeback. The critical question of when, if ever, the government would feel secure enough to lift its control was one that the current leadership was not prepared, and not called upon, to answer.

In the early weeks of 1949 the measure-for-measure cycle of Hungarian-American relations began, each indignity being answered in kind. On the surface the incidents were unconnected, but they were all symptoms of worsening relations. Washington retaliated for Chapin's expulsion by refusing visas to five Hungarian delegates intending to represent their country at a cultural and scientific conference in New York. The foreign ministry in Budapest thereupon charged that U.S. military attachés were illegally taking pictures in restricted border areas and asked for their recall. When the State Department refused, the officers were declared personae non gratae. As the Hungarian legation in Washington had no military personnel attached to it, the state department could not take appropriate countermeasures.[34] Chapin was succeeded by a chargé, William Cochran, who himself experienced the high-handedness of the Budapest government and asked in one of his reports home "how much of this disrespect and defiance, far exceeding Hungary's Soviet master's example, must we put up with from Hungary, which has obviously long abandoned not only all respect for international law, truth, decency, comity, but also seems to have lost all sense of proportion."[35]

The pattern Cochran perceived was unmistakable. The soviets had chosen Hungary for the thankless task of visiting on the United States indignities that the cautious soviet foreign affairs establishment thought best not to employ itself. The United States had played its principal trump card by suspending the return of Hungarian assets and now Hungarian foreign policy enjoyed an impunity it had not experienced since the time of the Dual Monarchy, if then. Budapest's subservience to Moscow did not diminish its sense of importance; it looked on itself, not as an agent of another country but as an instrument of international communism, a position that placed it above ordinary constraints. The officials who now came to direct the country's foreign policy succumbed to this heady sense of power; much of the diplomacy in the years that followed bore the stamp of an arrogance rooted in an obscure ideology.

Chapter 22

Widening Circles

A Hungarian government headed by a Smallholder had become a mockery of political realities. It was time for still another round of elections to effect a change. Mátyás Rákosi informed his contact at the Communist Party of the Soviet Union (CPSU) headquarters in Moscow, Mikhail Suslov, on April 13 that the parliament had been dissolved and new elections had been scheduled for May 15. The five coalition parties, as well as the trade unions, would run on a joint list under the name Hungarian Independent People's Front. "Our plan," Rákosi wrote, "is that 70 percent of the new [parliamentary] representatives should be members of our party, 20 percent from other parties should secretly belong to our party and only 10 percent should be undecided."

But who should assume the premiership after such an outcome? "In favor of a [communist premier] speaks the fact that the current state of affairs is anachronistic . . . conditions more and more demand a communist premier. Conversely, in favor of keeping our current premier is the fact that the struggle against the Catholic Church would be easier to wage with a non-communist premier. So would be the war against the kulaks and the introduction of collective farms."

At the end of his letter Rákosi mentioned in passing that a delegation from Czechoslovakia was due to arrive the next day to conclude a treaty of friendship and nonaggression. "This has become possible because, reluctantly though, the Czechoslovak comrades had begun treating the Hungarians in Slovakia better."[1] The actual fact was that the Czechoslovak comrades had begun treating Hungary itself better. Klement Gottwald prevailed on his politburo to agree to the cancellation of all existing financial obligations between the two countries, including reparations. Czechoslovak firms that had been nationalized in Hungary would be compensated for by corresponding expropriations of Hungarian firms in the republic.[2]

The rapprochement between these former antagonists was due in part to soviet pressure and in part to western hostility toward both. In the case of Czechoslovakia this hostility dated from the communist coup of February 1948; toward Hungary it had developed gradually. The communist leadership in Hungary accepted the worsening relations with equanimity; Rákosi, the effective head of government, was more concerned about the party affiliation of the next premier than with the state of relations with Britain and the United States. Just now, in the spring 1949, those powers, for unfathomable reasons, opted for a course of action that was practically guaranteed to fail and to prove embarrassing to them. After intensive consultations they decided to charge Hungary with violation of the peace treaty. The U. S. chargé, William Cochran, presented his government's démarche in that sense on March 10. In London the accusation was raised in the House of Commons.[3]

Then, on April 12, the two western envoys delivered parallel *notes verbale* to the foreign ministry. They called attention to Article 2 of the peace treaty, which obliged the Hungarian government to secure its citizens all human rights and freedoms, including freedom of expression, of press, of religion, and of political opinion. The British note, more specific than the American in citing instances of abuse of government power, ended with the sentence: "We find Hungarian government in violation of peace treaty and feel they should employ prompt remedial measures. Canadian, Australian, New Zealand governments, as they have no representation in Hungary, wish to associate themselves with above."[4]

The foreign ministry's reply was cast in the deliberately harsh and insulting tone that would become standard in communications with the western powers in the years to come. His Majesty's Government's (HMG) démarche, the ministry noted, was similar to the one presented at the same time by the American legation. The government was therefore led to conclude that "at the initiative of the U. S. Government, and agreed to by the Government of the U. K., a common diplomatic step was taken against the Government of Hungary."

There followed the virtuous assertion that Hungary had fully complied with the provisions of the peace treaty. In replying to the British charge that the judicial process had been perverted for political ends, the note referred to Article 4 of the peace treaty, "which explicitly obliges Hungary not only to dissolve . . . Fascist organizations but also not to allow the existence and activities of organizations of that nature." This was followed by a reminder that in Britain "serious discrimination exists between citizens of different races and colors and that, by a wide margin, not every person can enjoy human rights." In conclusion the note regretted the fact that "the Government of the U. K., having refrained lately from displaying an independent attitude, has joined the Government of the U. S. in its action against the Hungarian Republic."[5]

London and Washington, obviously realizing that nothing concrete would be forthcoming from this war of notes, referred the matter to the United Nations (UN). Against this move László Rajk on April 4 ranged the argument that since Hungary was not a member of this organization—precisely because of British and American objections—its actions were not subject to any ruling by that body, except if they imperiled international peace, and that was clearly not the case.[6]

The dispute was academic in any case, because even if the Mindszenty case was placed on the agenda of the UN, the Soviet Union would veto any adverse resolution. Nevertheless the question of whether to have a debate at all was brought before the UN Assembly on April 12. Thirty members voted in favor, seven against. A special commission was set up to decide whether Hungary should be invited to present its case; it voted heavily in favor of an invitation. When it was issued, Budapest turned it down. The foreign ministry reasserted its position that the Mindszenty trial, far from violating the peace treaty, had been held in its spirit, as it was directed against fascist organizations. The assembly then took up a motion to censure Hungary. The debate became mired in technicalities and was postponed to the fall.[7]

But while the government's pugnacity might have been a morale builder, it further increased the country's isolation. The great economic spurt that had driven reconstruction was leveling off. The Council of Mutual Economic Aid (COMECON), organized with much fanfare at Moscow's initiative, did not even come close to answering the vital needs of its member states. The intention that each country, instead of "modernizing" at a breakneck speed, should produce goods suited to its soil, climate, and expertise so that a healthy regional balance could be achieved might have been a felicitous one but ran into the opposition of the more backward states, who resented being relegated to "breadbasket" roles. Besides, even the purported industrial states lacked the capital to update their plants and their machinery and soviet investment remained nugatory and shoddy.

Hungary, as we have seen, had turned to Britain—or rather the British had expressed interest in economic ties after the conclusion of peace treaty, but politics interfered. American and Dominion pressures were no doubt at work but the sharp shift toward totalitarianism in Hungary was clearly the deciding factor. The May 15 elections produced an overwhelming electoral victory for the communist-dominated People's Front. A statement issued by the State Department on May 18 charged that freedom of expression had been denied, both in the campaign and in the election itself, and that "Hungarian Communist leadership had again drawn the world's attention to the totalitarian character of their regime."[8] On May 21 HMG resuscitated the debate over the violation of the peace treaty and asked for a three-power determination as to whether such a violation had in fact occurred, as provided by Article 40 of the treaty.*[9] On May 31 the British legation informed Georgy Pushkin that HMG had invoked Article 40 and asked for soviet participation.[10] Pushkin was in no hurry to reply and neither was the foreign ministry, which found itself in the throes of another change of personnel.

After the elections a new cabinet was formed in which László Rajk no longer had a place; unbeknownst to him he was under deep political suspicion. He was replaced as foreign minister by Gyula Kállai, a man of scant diplomatic experience, just at a time when the country's foreign affairs were becoming untidy. Even the new friendship treaty with Czechoslovakia was adding to the general confusion because it was concluded before the long-festering problems between the two countries were gotten out of the way. Signed in Budapest, the text of the treaty disingenuously stated that since the people now had taken control in both states, the conditions that had thus far militated against harmonious relations had disappeared.

The ratification debate supported the American argument that all oppositionist voices had been silenced. Premier István Dobi (who had succeeded Lajos Dinnyés five months before the elections), presenting the treaty to parliament, pontificated on another topic. "The more lasting becomes the

---

*Article 40 provided that in case of an unresolved dispute concerning the interpretation or execution of the treaty, it should be referred "at the request of either party to the dispute to a Commission composed of one representative of each party and a third member selected by mutual agreement of the two parties from nationals of a third country. Should the two parties fail to agree within a period of one month upon the appointment of a third member, the Secretary-General of the United Nations may be requested by either party to make the appointment."

solidarity of the people's democracies, among themselves as well as with the Soviet Union, the more glaring becomes the conduct of the shock troops of the imperialists to our south, of Tito and his accomplices. These recent deserters from the peace front seek to cause disturbances and provocations, spy networks, terror actions and the murder of our border guards; they are trying to retard our development."[11]

There was no "debate"; speaker after speaker hailed the treaty with Czechoslovakia. One speaker called it an important link in the chain of agreements among people's democracies, one that dashed the hope of international reaction that these two nations, divided by discord for so many years, would find it impossible to become peaceful neighbors.[12]

It had become a matter of common understanding that any attack on Josip Broz Tito and his "clique" was by extension an attack on western imperialism that had the Yugoslav dictator in its pay. Daily press reports spoke of border incidents in the south. On the consular level insults and counterinsults multiplied, followed by expulsions. Rákosi put his hand to an article entitled: "The Yugoslav Trotskyists, the Shock Troops of Imperialism," attacking Tito in immoderate language.[13]

When Rákosi spoke everybody listened, because his words were composed in Moscow. He alone stood above the infighting at every level of the party hierarchy, and even at foreign missions. At the time when the Tito expulsion caused suspicion to shift from reactionaries to communists themselves, Rákosi alone remained untouched by suspicion. However being untouchable was often an ephemeral state and he could not let his guard down or lower his vigilance. The time when he was ahead of Moscow in trying to bring the soviet system to Hungary had passed; the Tito affair had changed everything. The fact that Tito's quarrel was really with Stalin and not with the USSR or the people's democracies was of no further consequence; to be worth its name, a people's democracy had to make Stalin's position its own.*

(Yet on the operational level Hungary's relations with its mighty neighbor were not trouble free. Pushkin complained to Rákosi that certain members of the Hungarian Workers' Party (MDP) were displaying an unfriendly attitude toward soviet delegates in mixed enterprises and trying to hurt soviet interests. Rákosi readily admitted that the fault lay with Hungarian communists who allowed nationalism and antisovietism to find expression in the joint ventures. He added, with evident satisfaction, that at the same time sympathy for the

---

*Rákosi had expressed himself in this sense on the pages of *Szabad Nep*: "A people's democracy is a state that helps the Soviet Union to triumph and as such is supported by the Soviet Union; the laboring nation . . . progresses from capitalism to socialism."—Institute for Political History, Hungarian Workers' Party (MDP) Files, Central Leadership Documents, 276, 65/198.

Soviet Union was on the increase among workers and peasants. As to business enterprises, steps would be taken to remove the offending individuals.[14])

That a Titoist conspiracy would in time be discovered in Hungary too was a foregone conclusion, but that its central figure should be one of the oldest, most trusted, and popular members of the party was unexpected and shocking. Historians in Hungary are still debating whether the case against László Rajk was undertaken on direct orders from Moscow or whether Moscow provided only general guidelines while Rákosi and the political police did the rest. What is certain is that in Budapest Rákosi (together with Ernö Gerö, Mihály Farkas, and János Kádár) made the decision to arrest Rajk and then prepare a paper trail of treason, spying and, the crowning indignity: Titoism. This was the point in Rákosi's career at which he turned into a full-fledged Stalinist. Until now he had retained some of the traits of a true idealist—but there was no room left for idealism in the unremitting struggle against the imperialist enemy. In the USSR the practice of sacrificing a comrade who had shared with the leader the underground struggle and had given his essence to the party and the cause had long been a common feature of party politics; in the satellites it was a novelty and one that eventually undermined the position of every leader engaged in it.

Rajk was arrested on May 30 with six alleged accomplices. In addition to spying, treason, and Titoism, he was also charged with plotting to overthrow the Hungarian government. On the day after the arrest *Szabad Nép* published a brief noncommittal report; in the days that followed a slew of letters allegedly poured in from individual workers and workers' organizations, demanding swift justice and unsparing punishment for the traitors.

The Rajk case is a foreign policy item only because of its broader implications. The trial dossier, painfully collected over months, was highly eclectic, containing almost exclusively material that connected Rajk and his coconspirators to persons of a compromised reputation. The key figure was an American, Noel Field. The key country, by a quantum leap of association, was Yugoslavia. Field had been employed by the League of Nations until he became disillusioned with its ineffectiveness and went to work for the Unitarian Service Committee, a religious charity organization. The Committee was at that time involved in assisting republican volunteers in the Spanish Civil War to return home. It was largely thanks to these efforts and Field's direct help that Rajk had been able to return to Hungary in 1941; he at once went underground. It appeared that he was from the start an advocate of the "from below" model of national fronts, of which Tito was the chief exponent. As the Rajk file was being prepared, it was found that one of Rajk's close collaborators, Tibor Szönyi, had come back to Hungary from internment abroad with money provided by Noel Field, who in turn had received it from the Office of Strategic Services (OSS), then run by Allen Dulles. This fact conveniently established an "imperialist" connection. There was the additional fact that the repatriation of communists, mainly from France, during the war was mainly a Yugoslav operation. In the minds of the investigators it all hung together: American money, an American man-smuggler who was probably an imperialist agent, OSS involvement, and Yugoslav operational skills.[15]

As the net of suspicion widened, another dubious individual was caught in it. Lazar Brankov had come to Hungary in 1945 as a member of a Yugoslav military mission. As the bill of indictment explained, Brankov was Tito's personal agent, given the task to infiltrate the Hungarian Communist Party (MKP), produce internal dissolution, and popularize Tito and the Yugoslav political system. He received his instructions from Interior Minister Alexandar Rankovich, one of Tito's closest associates. It was the latter who ordered Brankov to establish contact with Rajk and prevail on him to promote the Titoist plan and eventually have the MKP top personnel, including Rákosi, assassinated.

After the expulsion of the Yugoslav Communist Party from the Cominform, so the indictment charged, Brankov was called home; he later returned to Hungary with instructions for Rajk to expedite the overthrow of the Hungarian people's democracy. Brankov, to deflect suspicion from himself, declared himself a political émigré, and condemned Tito and his policy. The Yugoslav government demanded his extradition on some trumped-up embezzlement charge in order to make his émigré status more credible. When Brankov's true mission came to light, he was arrested. In his alleged confession he implicated a score of other traitors besides Rajk and Szönyi. Having come this far, the Rajk case accomplished its mission: to prove the existence of an imperialist conspiracy within the MKP with a key person (Rajk) who built up spy networks in several people's democracies.[16]

(The apprehension of all the suspects and the coaching to make them into credible witnesses took months. In the meantime another "isolated" spy case made its way through the labyrinthine passages of the political police, the ÁVO, as well as the diplomatic apparatus. The facts as presented by the ÁVO were too melodramatic to be believable and the British legation in Budapest, as well as the Foreign Office, could be forgiven for suspecting another frame-up. At the end of June border guards arrested a woman, a telephone operator at the British legation, a Mrs. Gyula Torbay, as she tried to cross the border without a passport. Her past history made this attempt to slip into Austria doubly suspicious. She had after the war gone to Austria with her husband, a former Arrowcross air force lieutenant, who joined the British spy apparatus there. After working at the censorship office in Klagenfurt, he and his wife, on instructions from the British Secret Service, returned to Hungary. Both received employment at the British legation, she as a "telephonist," he as a chauffeur.

Mrs. Torbay's immediate chief was one Wallace Harrison, a married man. During her interrogation the woman admitted having an affair with Harrison and said that it was he who urged her to leave Hungary, as the authorities were about to arrest her as a spy. He promised to join her abroad, divorce his wife, and marry her.

What followed became the subject of a diplomatic imbroglio. Harrison reported to his superiors a rather far-fetched tale. By his account he received a telephone call from a person unknown to him who said that if Harrison wanted to know his lover's whereabouts he should come to a designated place, Heroes' Square, in Budapest. As he told it, when he got there he was forced into a car and driven to a house where he was allowed to see the woman "in a fainting condition." His captors told him that it was in his power to rescue her if he gave them the names of Hungarian traitors who were in contact with the British legation.[17]

His Majesty's recently appointed Minister, Geoffrey Wallinger, credited Harrison's story and dispatched a stiff note to the foreign ministry. That same day Harrison and his wife left the country in a great hurry. The point man on the case at the foreign ministry was Andor Berei, an exceptionally obnoxious person who took a delight in humiliating officials of the western powers. He asked Wallinger how he could believe that Harrison would go to a suspicious rendezvous on the basis of a phone call from an unknown person. Wallinger admitted it had been a case of poor judgment. He also explained that Harrison had been sent home because his wife had become hysterical.

In the days that followed Berei, who knew the true facts of the case, repeatedly asked Wallinger to withdraw the protest note, for otherwise his government would have no option but to make the facts of the case public. Wallinger however was not intimidated. Early in September Berei received a promotion and his place was taken by Endre Sik, formerly the Minister in Washington. Wallinger meanwhile pursued the case with a bulldog persistence, repeating the charge that a British diplomat had in effect been kidnapped and threatened. When Wallinger himself was called home, he paid a farewell visit to Foreign Minister Kállai and the latter for the first time told him that Harrison had offered his services to Hungarian intelligence. But

Kállai's hope that the revelation would persuade the British legation to withdraw its protest note was disappointed. The foreign ministry was most likely waiting for just this opportunity. Now all the blame for the public gaining a glimpse of this sordid affair rested with the British government.

On October 6 Budapest radio broadcast "the lies" of "the spy," "the agent," Wallace Harrison. Two days later the actual conversation between Harrison and agents of the secret police was aired, although it was in places unintelligible. Harrison spoke in a breathy voice, in great agitation. "I don't want the gentlemen here to think that I am part of the capitalist system because I am not. . . . I was taken from school when I was a young boy, a very young boy of 15, because my father and mother could not feed me. But now I have fallen in love with that girl and she disappeared and I do not know where she is. . . . All I want to know . . . tell an answer. I will help you all I can. . . . I am in a position to help you, I will help you." Here Harrison was growing distraught and added, "I mean that from my heart, I like her, I love her, I will work for her, I will work in the field, I will work anywhere, I mean that, you must believe me when I say that, I mean that from my heart."

These romantic Anglo-Saxons! Not only romantic, foolish as well, to appoint a man of such unsteady character to a sensitive position and back him up when he got himself into trouble. The magazine *East Europe*, published in London, carried an article putting the blame not on Harrison but on the Foreign Office. The magazine charged that the foreign service obviously had no suitable policy for making diplomatic appointments to totalitarian countries; such candidates should be tested for nervous resistance and be thoroughly trained in methods employed by the political police; no person born in a receiving country should be allowed to serve in that country for any length of time. And if one was observed becoming friendly with a national of that country, he should be recalled at once.)[18]

When the Harrison case entered the public domain the Rajk trial had just been completed. The public was numbed and bewildered by allegations that spies and foreign agents had conspired to overthrow the still nascent republic. The atmosphere was heavy with foreboding. The trial, like Mindszenty's, had been open and four days after it ended the American ambassador in Moscow, Alan G. Kirk, commented on it to the State Department:

> Rajk trial appears as Kremlin's most determined and serious public effort since emergence of Titoism to deal with this cancer of the body politic of the Soviet-communist world, reflecting Moscow's serious and growing concern over this disruptive force which shows signs of undermining one of the most basic principles of "proletarian

internationalism", (i.e. Soviet imperialism)—absolute control of satellites and all communists, both home and abroad. Just as the 1937 purges were carried out in order to ruthlessly eliminate all Soviet opposition to Stalin, the Rajk purge is now directed at the mortal sin of nationalism, specifically in Hungary but with worldwide implications, which threatens Kremlin's rule and authority.

Of some interest is fact that charges of Trotskyism are now being hurled at Rajk, Tito and their ilk. While Trotsky was originally branded a "left" deviationist, and "nationalism" has generally been regarded by Bolshevik dogma as a "rightist" sin, Trotskyism gradually developed into a general epithet applicable to all former disciples of Leninism-Stalinism who broke with or opposed authority and view of Stalin, thus now adaptable to modern postwar heretics such as Tito.[19]

It was at this most inauspicious time that Hungary, together with 11 other states, renewed its application for membership in the UN. All these applications were denied, only representatives of the people's democracies voting in their favor.[20] On September 3 the United States and the United Kingdom introduced in the Security Council a motion to censure Hungary and Bulgaria (and, in a modified version, Rumania) for failure to respect human rights and basic freedoms. (In the case of Hungary there were hints of, but no direct reference to, the Mindszenty and Rajk trials.) The soviet member of the Council, Andrey Vishinsky, protested. He stated that while in the people's democracies religious exercise was free, it was not allowed to be directed against the state system. On October 11 he delivered an impassioned speech on the subject. "What is the substance of these charges?" he queried. "The accusers refer to the criminal trials. But the whole world knows that these trials do not infringe on the peace treaty provisions. The pertinent articles have not been conceived in such a manner that those crimes against the nation, the fatherland, the people . . . treason, spying, conspiracy and other crimes should go unpunished."[21]

The new Hungarian envoy in Washington was Imre Horváth, an old communist of working-class background who had received his political education in Moscow. He presented his credentials on October 17. In Budapest, Selden Chapin was replaced by Nathaniel Davis. But relations had deteriorated far beyond the point where a change of personnel could make a difference. Diplomatic contacts had been reduced to protests and counterprotests, charges and denials, creating a most unwholesome atmosphere for the conduct of interstate affairs. The British persevered in their attempt to form a tripartite commission to examine their charge that Hungary had violated the peace treaty, and they were supported by the United States. These two also accused Rumania and Bulgaria of denying their citizens political and religious rights. The new soviet ambassador in Budapest, Alexander Panyuskin, countered the British-American invitation to participate on such a commission by saying that since in the soviet view Hungary had fully complied with the demands of the peace treaty, there was no need to convene a commission.

The peacemakers had foreseen such a disagreement and provided that if any of the great powers refused to sit on the commission the parties at dispute (in this case Hungary and Great Britain) were to agree on a third country that would take the place of the dissenting power. This required Hungarian cooperation and it was not forthcoming. The foreign ministry adopted the soviet position that there had been no violation and hence there was no need for a commission. An internal memo took note of the fact that

> the government of the USA has not even attempted . . . to dispute the arguments and facts presented by the Hungarian Government, just as it had no reply to the arguments of the Soviet Government. The American Government is unable to reconcile itself to the fact that Hungary, in consequence of its liberation by the army of the Soviet Union, has become a free and independent state which desires to determine its own fate; [US Government] wishes to use every opportunity and every pretext to interfere in Hungary's domestic affairs and to derail Hungary's internal development against the will and the goals of its laboring people, in the interests of a small and vanished [social] stratum.[22]

It is a puzzle what London and Washington hoped to accomplish by pressing an issue that, even if it was resolved in their favor, would not yield concrete benefits. They had made the point that Hungary and the other accused states were under effective soviet control too often to require reiteration. HMG, brushing aside the arguments of the foreign ministry, denied that Hungary had the right "to arrogate to itself solely the interpretation of the Treaty of Peace to which it is itself only one party." Further, "The claim that the Hungarian Government has complied with Article 4 of the Treaty [calling for measures to dissolve all organizations of the fascist type] does not excuse breaches of other articles. Article 4 was not intended to be used as a cloak for the denial of the fundamental freedoms specified in Article 2, nor as a pretext for the suppression of all opposition to the rule of a minority."[23]

Soviet apprehension about being drawn into the dispute is in part explained by the provision in Article 40 that on a commission the majority vote would be final and binding. The pressure on the Budapest government intensified accordingly. On September 19 "in noon hours," as a foreign ministry memo stated, the British and American Ministers delivered new démarches. Although they differed in language, their substance was identical and contained nothing beyond the previously leveled charges. The legal expert of the foreign ministry saw no risk in ignoring the notes. "Other than rupture of diplomatic relations," he wrote, "the only means open to these powers is to propose that the U.N. censure [us] and to oppose our admission . . . as heretofore. We should therefore not reply to their notes."[24]

This was by no means the end of the matter. HMG drew the Canadian and Australian governments into the dispute and it was now their turn to charge Hungary with violation of the peace treaty. Each such complaint necessitated the convocation of a commission on which the soviets would be invited to participate; they would refuse and Hungary would be called upon to nominate a replacement, as the British and Dominion states would nominate theirs.

Relations were heading toward a deep freeze. In the meantime the arrest, and at times disappearance, of persons working for western legations, continued. Seldom in the past had the Anglo-Saxon powers been drawn into such humiliating dealings with a small country or been so helpless in extricating themselves.

Chapter 23

Diplomacy by Abuse

On September 26, 1949, the day after the sentences in the Rajk trial were handed down, the foreign ministry sent a long *note verbale* to Belgrade, claiming the trial had revealed that "the personnel of the [Yugoslav] legation in Budapest . . . has organized an extensive spy net and has given over the material so acquired to members of the English and American diplomatic missions engaged in espionage." The Hungarian government was impelled to conclude that the Yugoslav legation in Budapest was the center of a widespread plot to overthrow the Hungarian people's democracy.

"All this," the note asserted, "is unparalleled in the history of international relations. The facts that came to light at the trial can be termed only political banditry."[1]

The Yugoslav government was not caught off guard. It had published an Information Directory, denouncing the trial as a deliberate anti-Yugoslav provocation. It was all part of a soviet-inspired campaign aimed at the violent overthrow of the legitimate Yugoslav government. Despite the losses the nation had suffered during the war, partly from invading Hungarian forces, it had kept its postwar demands on Hungary to a minimum in the interest of friendly neighborly relations. "We did not intend to vitiate the political and economic position of Hungary as a vanquished nation; on the contrary, we wanted to aid her quick recovery." The article recalled how often and how effusively the communist press in Hungary had expressed gratitude to Josip Broz Tito and the Yugoslav Communist Party. How was it possible then to believe that the charges leveled against them had any validity?[2]

Such argumentation fell on deaf ears, not only in Hungary but in all the people's democracies. The tenor of the exchanges was becoming so vehement that there was speculation in several quarters that it was all preliminary to some overt, possibly military, action. Reports of incursions by Yugoslav forces into sovereign Hungarian territory lent weight to this theory. Such rumors, however, had a way of getting out of hand and producing consequences neither Budapest nor Moscow intended. The foreign ministry found it necessary at the end of October to send circulars to its 24 foreign missions, denying reports that the people's democracies were carrying out troop concentrations or that they had deliberately provoked border incidents. Such reports, the circular stated, were spread by "Gestapo agents, police informants and agents of western imperialist states who at present sit in the Yugoslav government and, under American direction, keep approximately half a million troops under arms." An additional propaganda item stated that: "Expenses [incurred by] this vast army greatly contribute to the misery and privations which the Yugoslav people are compelled to endure."[3]

On November 4 the Hungarian minister in Belgrade received several inquiries from French and American officials asking whether it was true that a break in Hungarian-Yugoslav relations was impending. The Minister vigorously denied this and stated that diplomatic relations were normal, "though we do not tolerate that Yugoslav diplomats in Hungary interfere in internal affairs [by carrying out] anti-state, anti-democratic, anti-soviet activities."

To the foreign ministry in Budapest Szántó reported that the atmosphere in Belgrade was growing increasingly tense. The legation was kept under surveillance and merchants had been prevailed upon to withhold goods from its personnel. When it was necessary to send a diplomat to the foreign ministry, he was treated with studied discourtesy. There was reason to believe that it was really Belgrade that aimed at breaking diplomatic relations.[4]

The friendship treaties concluded two years earlier with Yugoslavia had become mere pieces of paper. On September 28, in Moscow, Deputy Foreign Minister Andrey Gromyko advised the Yugoslav ambassador that his country's treaty to that effect was canceled. Two days later the Hungarian foreign ministry also served notice. In due course Poland, Czechoslovakia, Rumania, and Bulgaria all declared their friendship treaties with Yugoslavia null and void, although in no instance were diplomatic relations broken. The war of words reached such proportions that even the epochal event of the communist victory in China was all but overlooked by the communist press in its preoccupation with Tito's treason.

When the Cominform met in Budapest in the middle of November the Rumanian delegate, Gheorghe Gheorghiu-Dei, introduced a resolution titled: "Yugoslav Communist Party in the Hands of Murderers and Spies." It charged that under the leadership of Tito and Rankovich the Yugoslav party had passed from the already despicable bourgeois-nationalist phase to fascism. It was the duty of every good communist to attempt to overthrow it.[5] On November 16 the conference drew up a list (as if one had been needed) of aggressive imperialist powers and included Yugoslavia in it; the document charged Yugoslavia's Communist Party with conspiracy to restore capitalism and overthrow the people's democracies. Had the seventieth birthday of Joseph Vissarionovich Stalin—the great leader of the working people of the world, the wise teacher of mankind—not come at this time, the drumfire of anti-Tito propaganda would have continued to monopolize the media. As it was, accounts of advance preparations, of trains laden with gifts rolling into the Moscow railroad station, took up so much space and their delirious tenor was in such sharp contrast to the sheer venom of anti-Tito propaganda, that the latter deferred to the former.

Hungary's Moscow embassy (it had been upgraded from a legation earlier that year) lost little time in adopting its tenor to the new propaganda line. While Gyula Szekfü had headed the mission and the staff was generally neutral, the reports reaching the foreign ministry in Budapest were evenhanded and depicted unsparingly the shortcomings of the soviet system, the backwardness of the economy, and the slow pace of reconstruction. By the end of 1949 objectivity had fallen victim to the intense ideological pressure that the Two Camps concept and the campaign against Tito demanded. The new ambassador, András Szobek, a former carpenter, had been a member of the Red Guard during Hungary's failed revolution in 1919. He was, as we shall see later, no great admirer of Russian character or the soviet system. Perhaps in part to compensate for that, and also because no ambassador could for long keep the Moscow post unless he uncritically endorsed soviet policies and achievements, his reports brimmed with adulation and obsequiousness. His summary of the year 1948 had been late because of his ill-timed replacement of Szekfü but made up for this with an almost youthful enthusiasm. "The USSR has been strengthened as the people of the world became aware, as never before, of her leading role [in the fight] for progress and socialism and global peace. . . . Economic success is outstanding proof of decisive superiority of socialist state system, of the unity of the soviet people and the measureless dedication with which the people of the socialist empire close ranks behind the Bolshevik party and its great leader, Comrade Stalin."

As for Hungarian-soviet relations, "Especially heartening is the fact that under the impact of the Mindszenty and Rajk trials soviet public opinion and the whole attitude of the soviet people has decisively changed to our advantage. Their former reserve, which could be felt for a long time after liberation, has disappeared; we now experience sympathy and the desire to help."

In reviewing cultural developments, Szobek succumbed to the current vogue of involuted Marxist jargon that effectively obscured meaning, as it was probably intended to. This was the time, it should be remembered, when Stalin imposed on all creative expression an intense nationalistic stamp and a rejection of all foreign elements and influence. The struggle in this direction, Szobek wrote, was now targeted against a group of drama critics (but extended to other groups as well) "who judge features of socialist culture from a formulistic, idealist, decadent bourgeois viewpoint, [but] are unable to fit them into the bourgeois esthetic-scientific norms and therefore reject them altogether. In the final analysis it is a matter of removing the remaining opposition to already formed and clearly delineated conceptual directions as laid down by party directives."

Commenting on this part of the report the reader in Budapest observed that it was thoroughly and scientifically presented and had as its point of gravity the most important aspects of soviet cultural life. "Of the entire report this section is the most thorough and most valuable," he concluded.[6]

To a politically more sensitive reader the earlier paragraph, which linked growing sympathy for Hungary to the staged political trials, would have been far more instructive. The primacy of political correctness over all other areas of endeavor has never been more evident in Stalin's long regime. Nowhere does Szobek's report have anything to say about soviet recognition of Hungary's impressive economic recovery and its becoming the most prosperous member of the east bloc; just as in 1947 the elimination of Béla Kovács and Ferenc Nagy had served as proof of Hungary's commitment to the peace camp and endorsement of soviet ideals, in 1949 the removal of a cardinal and the trial and execution of a dozen faithful communists earned the highest appreciation of the Kremlin and the soviet people.

With the emphasis on searching out espionage, treason, and sabotage, foreign citizens in the people's democracies found their scope of activities, as well as their personal freedom, ever more restricted. When an employee of a British or American enterprise tried to salvage what he could from a nationalized firm, he was often accused of some nefarious design. In the

immediate postwar years when the country desperately needed foreign assistance and the political climate was more benign, authorities generally
blinked at such activities. But the economic crisis had since been weathered; much of what in the past had been looked on as normal business activity was now branded capitalistic manipulation and the position of western
businessmen became ever more precarious. This is not to suggest that some
of the charges brought against these men were not well-founded or that
their harassment and prosecution occurred wholesale. Hundreds of businessmen conducted their affairs undisturbed and were allowed to enter and
leave Hungary as they pleased. The charge made by the American Minister
that people's democracies "always assume foreigners are spies,"[7] was certainly a gross exaggeration. However if a man of affairs represented a major
corporation that, as the communists saw it, had political clout in its home
country, his activities were monitored with particular prejudice.

The case of Robert Vogeler, an American citizen and a special representative of the International Standard Electric Corporation in Hungary,
Austria, and Czechoslovakia, was a case in point. He was arrested by the
security police ÁVO on November 18, during a trip from Budapest to Vienna.
No official announcement of the arrest was made and, so far as the American envoy Nathaniel Davis was concerned, Vogeler simply disappeared.

An interview with Foreign Minister Gyula Kállai produced no results.
The fact was that Kállai depended for any information on the political police and none was forthcoming. Reporting to the State Department, Davis
wrote, "Police statement to Foreign Minister denying knowledge [of what
happened to Vogeler] must be as unsatisfactory to Foreign Minister as to
me." In a subsequent audience Kállai admitted that Vogeler was under arrest but when Davis requested information on the charges against him and
permission for the American consul to visit him, Kállai replied lamely that
Hungarian law did not permit giving such information or allowing such a
visit until the investigation had been completed.[8]

Meanwhile another businessman, a British citizen named Edgar Sanders, had also been arrested, allegedly on the basis of information given by a
Hungarian associate, Imre Geiger, a director of the Budapest branch of Standard Electric. Geiger had been taken into custody ten days earlier as he tried
to leave the country illegally.

These arrests marked the beginning of a protracted and, for the western
powers, frustrating diplomatic interplay that led to a catastrophic worsening of relations between Hungary on the one hand and Britain and the United
States on the other. Kállai's demeanor during his conversations with the
western envoys made it amply clear that he lacked authority not only to

make decisions, but even to provide information unless he had been expressly authorized to do so. Davis noted in one of his dispatches that Kállai "appeared nervous whether this indicates anything specific I don't know."[9] As soon as the true nature of the case became known to him, Davis requested an interview, not with Kállai, but with the man who had all the authority the foreign minister lacked and who was the ultimate judge of sensitive cases. But Mátyás Rákosi, on various pretexts, avoided meeting with Davis; at one point a date was set for an interview but it was later canceled. Kállai excused the cancellation by saying, untruthfully, that Rákosi had gone on vacation.

The American envoy was from this point on reduced to a humble petitioner, although his communications to the Hungarian foreign ministry bore every sign of dignity and gravity. They were however sent into an apparent vacuum. The British minister, Alexander Knox Helm, did not have better luck either. Vogeler and Sanders were being held incommunicado and all attempts to visit them were rebuffed. In London, the Foreign Office was considering suspending the trade talks with Hungary if the impasse continued.[10] That prospect failed to move the Hungarian Workers' Party (MDP) leadership in Budapest. It had found a new, if rather discreditable, way of leading from strength and the full possibilities still had to be tested.

In the midst of these exchanges still another American citizen, Israel Jacobson, director of the JOINT Distribution Committee (a worldwide Jewish relief agency) in Hungary, was arrested on charges of espionage. The by now familiar cycle of protests, evasions, and rebuttals was reenacted and every sign pointed to a new sensational espionage trial—then Jacobson was unexpectedly taken to Vienna and released. Possibly the intent was to demonstrate that each "spy" case was treated on its merits—or perhaps the prospect of the JOINT discontinuing its philanthropic activities in retaliation gave the communist leadership second thoughts. In any case, the political damage had again been done.

The price Hungary had to pay for this deterioration of relations with the western powers was high: it was cut off from profitable investments and foreign aid. This of course might have been deliberate because it increased the possibilities of cooperation among the communist states united in the Council of Mutual Economic Aid (COMECON), established to make the soviet-bloc economically self-sufficient. At this stage however economic questions were still of secondary significance; the political soil had to be hardened first. Hungary's battle of wills with the western powers, in which it had the full support of the Soviet Union, gave it an illusion of strength it had never experienced under previous regimes, when the nation lacked a unifying ideology.

Chapter 24

Points of Friction

Hungary had to pay a price, not only for antagonizing the western powers, but also for joining with other soviet-bloc countries in condemning and ostracizing Josip Broz Tito's Yugoslavia. A trade agreement made between the two countries the year before was about to expire and Belgrade evinced no interest in renewing it. Hungary faced the prospect of losing large quantities of timber and iron provided for in the agreement as well as the disappearance of Yugoslav markets for its finished products. A study prepared in the Hungarian Workers' Party (MDP) secretariat estimated that the country would be left with $5.1 million worth of finished goods that it could not sell anywhere else. There was also the possibility that Yugoslavia would close the harbors of Rijeka and Trieste to Hungarian goods. In such a case, the study stated, talks would have to be initiated with Poland for the use of Gdynia harbor and with Rumania for sending goods through Constance on the Black Sea. The USSR might be persuaded to make up the loss of timber and Czechoslovakia the loss of iron after the lapse of the treaty with Yugoslavia.[1]

In spite of these difficulties the tenor of diplomatic exchanges did not soften. A note sent to Belgrade on June 16 placed the blame for the worsening trade relations and the narrowing of the annual exchange of goods from a $12 million value to $8 million on the Yugoslav government. The note charged that even on this diminished scale Yugoslavia was in arrears on deliveries for the years 1947 and 1948. All this was proof that it was either unable to pursue a planned economy or that it was hostile to the nations advancing on the socialist path. The note referred to another existing trade agreement and concluded that under the circumstances it was defunct. Long-range agreements were possible only among countries with socialist economies.[2]

Although in its content the note limited itself to economic matters, it had an unmistakable political edge that Belgrade did not ignore. The Yugoslav government nevertheless waited five months with its reply, which bristled with reproaches. It reminded the government in Budapest that Hungary had unilaterally canceled the 20-year friendship agreement that the two countries had concluded in December 1947. This was all the more regrettable as it was the first such agreement Hungary had made with a neighbor state. The Yugoslav démarche further noted that only seven months after the conclusion of that agreement Hungary had been the first to join "certain soviet leaders" in an intemperate attack on its southern neighbor. With this, it seized the opportunity "to plant the seeds of chauvinistic hatred and revive the St. Stephen idea against Yugoslavia."

This broad charge was followed by a list of particulars, chief among them that Hungary had staged "a huge provocative criminal trial [which was] based on inventions and fabrications and exposed the shameful endeavor to introduce new criminal methods into relations among nations. . . . It has caused immeasurable harm to the cause of peace and [in the end] ruined its perpetrators." (Gyula Kallai sent a translation of the note to Mátyás Rákosi with the handwritten comment: "We will examine it and clear with other interested states whether we should answer it. I don't see much sense in it.")[3]

Perhaps some of the loss resulting from the cessation of trade with Yugoslavia could be made up by an agreement with England. Negotiations had been going on for a while but now the arrest of Edgar Sanders and his being held incommunicado put those too in jeopardy. On December 19 Geoffrey Wallinger visited Andor Berei (who had again been put in charge of dealing with the western powers—apparently Endre Sik was not forceful enough). Wallinger delivered still another diplomatic note warning that trade talks might suffer if the British consul was not allowed to visit Sanders. Berei contemptuously referred to this note as *"un bout de papier"* (a piece of paper). Wallinger returned the next day to say that the *bout de papier* had expressed his personal opinion but that he had now heard from London and was instructed to deliver the following message: "Until Mr. Sanders had been visited by the British consul in conditions enabling them to talk in privacy and until HMG are satisfied that Mr. Sanders is receiving proper treatment, all trade and financial negotiations, including negotiations for the purchase of British shareholdings in Hungary, are suspended."

The Hungarian government delivered its reply in a public note so to speak, by printing it in a foreign ministry communiqué on December 19. It recapitulated the English threats and stated that the head of the Hungarian commercial delegation in London had already advised the English side that inasmuch as it had made demands contrary to the original terms, his delegation would return to Budapest. The communiqué went on to charge that the English note was an attempt to influence Hungary's treatment of a spy. The Budapest government categorically refused to make the future of its trade relations dependent on the outcome of an espionage case.[4] (In an internal memo three weeks later Berei recorded his answer to Wallinger's communication as stating that commercial ties were as important to His Majesty's Government as they were to Hungary and thus threats were uncalled for.)[5]

The cases against Sanders and Robert Vogeler were in the hands of the ÁVO chief Gábor Péter. Rákosi kept a tight control over all preparations. He had reviewed the transcripts of Vogeler's interrogations and sent Peter a list of items that required clarification. "Was English spying subordinate to the American?" "Data on Vogeler's previous spying, his past as a spy, is missing from the minutes." "We have to check the names of all the diplomats [involved in the case] so that no liar will remain among them."

In a separate communication Rákosi demanded that not later than February 5 (the trial was scheduled to open on February 17) the entire timetable of the judicial procedures had to be prepared, just as it had been in the Mindszenty and Rajk trials: how long the questioning of each defendant would last, how much time the summations would take up, and how soon after their completion the verdict would be announced. The indictments had to be specific about how long Sanders and Vogeler had been spies, what schools they had graduated from, and, in the case of Vogeler, in which factories he had organized a system of spying on workers. The greatest care had to be taken that the accused would not retract their confessions. Vogeler had to be asked to give the names of high-ranking officers, generals, and admirals, who sat on the board of Standard Electric, so as to demonstrate the close links between the armed forces and American capitalists. Vogeler would have to confess that his prior mission, in Vienna, had the full approval of the Joint Chiefs of Staff. He was also expected to confirm that most (if not all) Americans going abroad were engaged in intelligence work and that on their return home they reported to the competent spy organizations. Finally, he was to admit that American enterprises in Hungary were, without exception, engaged in espionage work.[6]

On January 19, after weeks of pressing for an interview with Rákosi, Nathaniel Davis was finally granted one. It proved a disappointing occasion. Rákosi turned away the envoy's plea on Vogeler's behalf; he said his government was determined to try the American as a spy and prove its case. The evidence would reveal that the American diplomatic establishment was closely linked to Vogeler's activities. The responsible persons, Rákosi added, "can expect the usual treatment of diplomats who overstep their legitimate duties."[7]

All this time, for no other reason than its nuisance value, the British and Americans kept the matter of Hungary's alleged violation of the peace treaty alive. They periodically advised the foreign ministry in Budapest of their choice of persons to sit on the investigative commissions in lieu of the Russians, and urged Budapest to announce its own choices. The United Nations General Assembly, which had tabled the case in its spring session, had requested a ruling from the Hague International Tribunal on three specific questions: (1) Was there a legal dispute between Hungary on the one hand and the United States, Great Britain, and the Dominions on the other? (2) Was Hungary obliged to name representatives to the commission? (3) Did the secretary general of the UN have the right to name such delegates if the Hungarian government refused to?

The Hague Tribunal answered each question in the affirmative. The General Assembly, on October 22, invited Hungary (as well as Rumania and Bulgaria) to name its representatives. Budapest reiterated its position that the circumstances requiring the appointment of such a commission were not present.[8] The General Assembly could at this point have instructed the secretary general to make a selection but, probably realizing the futility of a commission on which neither Hungary nor the Soviet Union was represented, refrained from doing so. The western powers had only the meager satisfaction of showing to the world that the soviet satellites had no respect for international forums and international practices.

They cited the cases of Vogeler and Sanders as proof of that contention. It was a commonly respected right of a person taken into custody in a foreign country to be visited by his consul and to confer with him in conditions of privacy. Berei kept referring to the Hungarian law that forbade such visits while the investigation was pending. The question of whether the laws of the country within its area of jurisdiction superseded those of international law was again one that no court could satisfactorily and bindingly answer.

Wallinger kept pressing: could Sanders be represented by a British lawyer? When Berei refused, Wallinger asked whether a lawyer might act in an advisory capacity. Berei advised Wallinger to study the Hungarian law code and discover whether such indirect representation was permitted. Wallinger reported on February 11 that he had found no prohibition against it. He wrote to Berei afterward: "While you did not receive my above communication with great glee, my plea that justice be done may have made some impression on you. The name of the lawyer [who had offered to participate in the Sanders trial] is Mr. Griffith Jones . . . even in Hitler's time Dimitrov was allowed to have a Bulgarian lawyer's assistance in addition to the German one. Please issue Griffith Jones a visa."

Of Berei's reply only a rough draft had been placed in the files and it is undated. It refers to previous communications that made it clear Hungarian law did not permit participation of foreign advocates in criminal trials. "I also find it peculiar," Berei wrote, "that you refer to Dimitrov's precedent in this case and give the Hungarian People's Republic Hitler's Germany as an example."[9]

The commercial delegation meanwhile was cooling its heels in London. It spent its time composing an analysis of the British political and economic scene in the global setting. The crisis of capitalism, the study stated, was becoming more profound. Capitalism was concentrating its remaining resources on postponing a breakdown. In Great Britain two opposing forces were at work. One was a growing dependence on the United States in political and economic matters. Counteracting this tendency was Commonwealth solidarity. But, under American pressure and the dissolution of the sterling bloc, this solidarity was becoming meaningless. The year 1949 clearly had shown that it was no longer possible to speak of an independent English policy; every phase of it bore the stamp of dependence on the United States. In its imperial policy too Great Britain faced American opposition at every step. Domestically, it was American help that kept the Labour Government in power. This produced the paradoxical situation that seemingly socialistic measures were being made possible with American aid.[10]

The curious thing about this document was that its secondary premise canceled out the primary one. England's dependence on the United States, although somewhat exaggerated, was a fact, but it had nothing to do with the global crisis of capitalism—rather it confirmed the commanding power of its very citadel, the United States. The law of unequal development, postulated by Lenin, had a reverse application as the United States, a newcomer on the industrial scene, took advantage of the sweeping unrest in the colonial world and succeeded, with its neocolonial policies, in displacing the old, tired imperialist powers. The Soviet Union tried to respond by posing as the champion of national liberation movements, but in a world hungry for technology and investment, its ideological weapons proved puny and ineffective.

The trial of Sanders and Vogeler and their codefendants took place between February 17 and 20. They were all found guilty as charged—two Hungarians were sentenced to death; Vogeler received 15 years at hard labor, Sanders, 13 years.

The government used the alleged revelations of the trial to demand the recall of parts of the American and British legation personnel. Undeniably, the size of the staffs was too large (especially when compared to those of the Hungarian legations in London and Washington), but the demands again had a vitiating political content. As Davis put it in one of his dispatches: "Having publicly and repeatedly asserted USA legation is mere spy organization, any agreement as to size of staff they will tolerate will mean only we are here on sufferance, branded as spies but allowed to remain so long as by their dictum innocuous ones."[11]

Hungary had by now absorbed the loss caused by the sequestration of its assets in Germany and Austria and their bargaining value had accordingly diminished. Its exclusion from the UN was also a settled matter and the western powers had no more means of reward and punishment left. And so the government continued to challenge the capitalist states, not only with staged trials and gratuitous expulsions but in the economic sphere as well. The Nationalization Law of December 1949 mandated the seizure by the government of all the shares owned by foreign governments or foreign corporations in industrial and transport enterprises. There followed another round of futile exchanges. When Washington protested, the government in Budapest pointed out that it had expelled members of the Hungarian restitution commission from the American occupation zone in Germany. It was the first time that Budapest hinted at a quid pro quo settlement of its claim to sequestered assets abroad. Washington responded by closing the Hungarian consulates in New York and Cleveland, on the argument that the U. S. consul in Budapest was not allowed to carry out his official duties (i.e. visit Vogeler in prison) and thus the Hungarian consuls should be similarly restricted. The only effect of the closures was to put an end to what the Budapest government was not anxious to perpetuate anyway, contacts between its consular personnel and Hungarians well established in the United States.

(Rákosi had a private letter from the secretary general of the World Federation of Hungarians, complaining of just that rupture of contacts. "Our people's democracy acts like a magnet on progressive Hungarians [abroad]," the secretary general wrote. "It is the proletarian fatherland which they had learned to love in the workers' movement and for whose coming they have labored. Many would now like to go home, they write to us, we forward their letters to the KEOKH [Central Authority for Control of Foreigners— in Budapest], and their letters get lost in the bureaucratic quagmire. According to [your] officials, [many of the] requests are prepared in improper form, are set aside and not answered. At the same time impostors . . . hire agents and manage to obtain visas that honest comrades fail to get. Could you not intervene? Many of those who did return and applied for membership in the party were told that there was a freeze. Can't you make exception in such cases? How do we explain to comrades who had spent years in the workers' movement that they are not wanted in their own party?") (There is no trace in the archives of an answer to this letter.)[12]

Vogeler's conviction did not end official interventions on his behalf. He was still kept in isolation, though by now the argument that the investigation had not been completed had fallen away. Nathaniel Davis was instructed by the State Department to seek out Rákosi and he did so on March 27. He cited a new precedent (the one referring to Georgy Dimitrov had proved embarrassing) of a soviet agent who had been arrested in the United States, then released upon the intervention of soviet Deputy Foreign Minister Andrey Vishinsky. Rákosi pointedly remarked that the United States was asking for a great deal and was offering little. It was the first hint that Vogeler's release might be obtained in exchange for material concessions and it would be followed by a long round of cynical bazaar haggling. The Hungarian government presented an ever-escalating list of demands to which Washington agreed in little installments. On April 11 Berei (the true power in the foreign ministry) summoned Davis and asked him what the United States was prepared to do in still unsettled matters such as the status of the Hungarian consulates and the question of the release of assets. Davis reported to the State Department: "Hungarian Government inclined comply our Vogeler proposal but wished to hear concrete offers."[13] (Actually no proposal had been made, only a demand for his release.)

The State Department took the position that while it might lift the retaliatory measures taken after Vogeler's arrest, other pending issues would not be part of the give-and-take. Only in the matter of the displaced goods was there some flexibility; once Vogeler was released, the U. S. government would "facilitate delivery all Hungarian goods in US zone Germany which have been found available for restitution."[14] However the criteria for availability were not stated and the goods declared available were of minor value—by the department's estimate $360,000.

Several times during these negotiations it appeared that, all obstacles having been cleared away, Vogeler's release was imminent. But always new difficulties arose, either because of some indiscretion or because the Hungarian side perceived opportunities for greater American concessions. The Voice of America (VOA) broadcasts were an ever-recurring topic. Once, when the conditions of Vogeler's release had in effect been settled, the news of the secret talks was leaked in Vienna (probably by a British legation officer) and the release was canceled. When the talks were resumed Berei introduced the question of the Crown of St. Stephen. Later he declared himself willing to drop it in exchange for the complete cessation of VOA broadcasts. It was now the end of June, Vogeler had been in prison for six months, and Dean Acheson finally put his foot down. "Hungarian demand for abandonment of VOA Hungarian-language broadcasts from Munich as quid pro quo Vogeler release entirely out of the question," he wrote to Davis.[15] Berei had overplayed his hand. With the talks at a deadlock, no concessions at all were forthcoming. For now Budapest had only the crumb of satisfaction of keeping an American businessman prisoner.

His Majesty's Government (HMG) also stiffened its stance, for a more concrete reason. By early 1950 the harassment of British diplomatic and nondiplomatic personnel had become so common that much of the legation's activity was taken up investigating the various incidents and negotiating their settlement. In March a legation secretary and a military attaché were declared personae non gratae. On March 25 the legation, on instructions from London, conveyed to the foreign ministry the following message: "HMG has no option but to comply with requests that Messrs Southby and Capron be recalled. But it rejects grounds for request. Case against them rests entirely on statements made by Messrs Sanders and Rado [one of Sanders's alleged Hungarian accomplices] in connection with former's trial. . . . HMG totally unable to accept allegation that Lt. Col. Capron and Mr. Southby were guilty of actions improper for diplomatic officers."[16]

Such communications were routinely left unanswered. So were protests over the arrest of Hungarian employees of the legation. The foreign ministry kept up its complaints over the excessive size of the legation staff. Disputes became ever more petty and petulant. A small but typical example was a foreign ministry note to the ÁVO, alerting it that a street vendor on a busy square was selling fruit in bags pasted together from a British legation bulletin: "we think it is likely that we deal with clever imperialist propaganda and request that proper measures be taken in this matter."[17]

The expulsion of diplomats accused of overstepping the scope of their official assignment was common, but because it was done wholesale the government opened itself to the charge of violating rules of comity in international relations. The foreign ministry generally declared itself indifferent to such charges but their effect was cumulative.

With the mood outside the peace camp turning hostile to it, the government decided to present its own case to the Hague Tribunal. Dated January 13, 1950, the brief argued that neither the United Nations nor the Hague Tribunal was competent in deciding whether Hungary did or did not violate the peace treaty. Through no fault of its own, Hungary was excluded from the UN. While it was true that the UN Charter stated that the UN might act when international peace and security were threatened, even against a nonmember, "Hungarian Government, far from [doing that] has in fact consolidated peace and security." Thus the UN acted improperly when it referred the matter to the Hague Tribunal. Nevertheless, the note stated in closing, the Tribunal had adopted, "without any further reasoning," the position presented by the powers that had brought those charges. Hungary therefore felt free not to comply.[18]

Chapter 25

The Embassy File

There is no logical and only a loose chronological sequence to this selection of legation correspondence from such diverse posts as Moscow, London, Paris, Washington, Belgrade, and Bucharest. It runs the gamut from the professionally analytical to the petty and frivolous and is intended to give the reader a feel of diplomacy on the operational level where the concerns of high politics are only distantly reflected.

We have so far had very little to say about Hungarian-French relations; they were sporadic and generally insubstantial. Early reports from Paris by the Hungarian mission there do not qualify for diplomatic dispatches as they were sent by people lacking an official status and found their way into foreign ministry files by default rather than intent. One, dated October 18, 1945, mentions a Hungarian consulate in Paris manned by personnel left over from the wartime regime and operating without an exequatur from the French government. When a Hungarian trade union delegation arrived in Paris about this time, a French foreign ministry official told its head, István Kossa, that his government was interested in establishing formal relations with Hungary and suggested that the Provisional Government approach the Allied Control Commission (ACC), or its French contingent, in the matter. The consulate in Paris, the official said, did not represent the Hungarian government and should be closed.[1] (This report came from a left-wing Hungarian living in Paris and is most likely biased; the French could easily have put that leftover consulate out of business if they had so desired.)

Kossa in his report said that his trade union delegation used its presence in Paris to discredit rumors of undue soviet influence over Hungarian politics. In a separate section he noted that the consulate in Paris, which had been appointed by the Nazi-dominated regime, was issuing Hungarian passports to a number of fascists who were infiltrating from Germany. The consul-general was hosting dinner affairs for an assortment of reactionaries, among them the Archduke Charles of the House of Habsburg, younger brother of the pretender, Otto. An agent of the consulate was furthermore regularly visiting prisoner-of-war camps and agitating against the current Hungarian government. A priest held captive in one of those camps was recruiting volunteers for the Foreign Legion, describing conditions in Hungary in such appalling terms that prisoners preferred to join the Legion rather than return home.[2]

The exact date at which Hungary and France established diplomatic relations is uncertain. A memo by a French foreign ministry official visiting Budapest, dated January 18, 1946, noted that a French "member" (of the ACC?) appeared at the Hungarian foreign ministry saying he had received a telegram from Paris to the effect that the French government had recognized the Provisional Government of Hungary and was ready to take up diplomatic relations; he asked for an *agrément* for a Henri Gauquier as Minister.[3]

(An undated and unsigned foreign ministry memo addressed to the cabinet states that the news organ of "a coalition party"—almost certainly the Social Democrats—charged that Paris had since the fall of 1945 wanted to take up relations but five months later it had still not named an envoy because the Hungarian government had made no provision for an apartment and an office. The paper demanded that the responsible officials be brought before a people's court.[4] In the meantime some official Hungarian presence must have been established in Paris because on February 9, 1946 it sent home a report that the Quai d'Orsay had informed it that the Czechoslovak government had protested against the French intention of taking up relations with Hungary, calling attention to the fact that none of the great powers had yet done so. Such a claim, if indeed made, must have been prompted by peeve, as Prague was well aware that by now both the United States and the Soviet Union had established relations with Hungary. The French foreign ministry accordingly asked the Hungarian mission for the exact date when Georgy Pushkin and Arthur Schoenfeld respectively had presented their credentials.)[5] (The first Hungarian Minister in Paris was Pál Auer, a diplomat of the Old Guard, who later resigned in protest against Ferenc Nagy's ouster.)

With Great Britain relations were also unsettled. On September 19, 1945 an interparty conference in Budapest, chaired by premier Béla Dálnoky-Miklós, was shocked to hear that the legation personnel in London was still the same as it had been under the Horthy regime. Dálnoky-Miklós offered this defense: "Admittedly, they [the legation people] have no contact with Budapest and hadn't had any for years, but it is certain that whatever they do is the right thing." He added that the diplomats still at their posts had been recalled for reaccreditation to determine whether some of them might be useful to the new regime.[6]

In Communist Party circles there was great concern about the character of the diplomatic establishment. An internal memo noted that the old corps of diplomats was clubbish, an elitist aggregation of wealthy men distinguished in appearance, financially independent, speaking a number of foreign languages. Although they were a socially anachronistic breed, the country still needed them to train a new generation of diplomats. But care had to be taken that the key positions in the foreign ministry were given to communists and that the foreign service was kept under close control. Nepotism, the evil of the past system, had to be eliminated. At the same time the level of performance had to be maintained.

In another place the memo noted that lately there had been a lot of talk about concepts such as "United Europe," "Danubian Confederation," and the like, especially in Anglo-Saxon circles. While these might have some transitory value, communists must never forget that their program was a global socialistic society. "Marx and Engels would turn in their graves if they heard the phrase, 'Proletarians of the continent unite!'"[7]

In this first postwar year relations with Moscow were probably the smoothest, although the Minister, Gyula Szekfü, might not have been the best choice for the post. While his reports were superficially prosoviet, they always hinted at doubts about soviet intentions. When for instance he hailed the purity and steadfastness of the soviet constitution he added that one of its basic principles was that the USSR had the right to aid any nation defending its freedom and independence. The envoy obviously knew very well what that aid meant in practice.

In referring to latent fears that the Soviet Union intended to incorporate its European neighbors into the soviet state structure, Szekfü wrote that, "apart from principles," it was not "at present" in the interest of the Soviet Union to annex border states; that would impose a grave economic burden on the soviet state. On the contrary, it was in the Soviet Union's interest that those states enjoy an independent existence, as long as their relations with the USSR are correct and neighborly.[8]

In one of his early dispatches Szekfü noted that despite the terrible destruction visited on it by the war, the USSR still had immense natural resources that enabled it to aid countries on its borders. Its reserves of items like grain, cotton, sugar, iron, and crude oil were such that if it threw them on the world market the effect on the western powers would be disastrous. But the soviets strove for good relations with those powers. They were also anxious to live on the best possible terms with the nations on their borders. Unfortunately Hungary was not looked on as a friendly neighbor. To cite one example, Moscow had offered to Hungary a gift of 250 trucks, to be delivered in the USSR. The Budapest government, instead of sending a devout communist to receive the gift, sent a count and a university professor. That made a bad impression. There were also antisoviet articles in the Hungarian press and demonstrations by university students. If Hungarians starved that winter, they would have only themselves to blame.[9]

Moscow was the most prestigious diplomatic post at the time; it was also a place of mystery, physical decrepitude, and rather odd business hours. Diplomats had to adjust their hours to the schedule of the foreign ministry. Mornings the ministry was staffed by only auxiliary personnel; no high functionary was at his desk before 1 P.M. Serious business began to be transacted only at about 3 P.M. and appointments had to be made between those two hours. On the other hand the ministry stayed open late. Telephone calls by an important foreign embassy official were often answered as late as 1 A.M. Business among embassies however was conducted during the daytime hours and the staff had to work all day and late into the night in order to accommodate both the foreign ministry and fellow diplomats.

Means of communication in Moscow were cumbersome and the mail was extremely slow. The Scandinavian states were helpful in placing their courier services at the disposal of other legations. The largest diplomatic mission, in the middle of 1946 was the American, with 38 staff members; followed by the British, with 27; the Chinese, with 23; and the French, with 16. At legations of former belligerent countries the "new element" was preponderant. At the Finnish mission, except for one secretary, the entire personnel had started its diplomatic career at that post. Most diplomats spoke Russian fairly well and in general, Russian was the language of routine contacts and negotiations.

Accommodations were generally miserable. Only the diplomats of the three western powers and of Poland were adequately housed. Most missions occupied hotel suites; offices and private quarters were often combined. The personnel of even those embassies who had their own buildings was usually quartered in hotels.

The everyday life of diplomats was dreary. There were very few places of public entertainment. Moscow was a stone desert with little green in the streets or the squares. Sports opportunities were virtually nil. A person had to go out of town to breathe fresh air. Prices of ordinary commodities were forbiddingly high. However the embassies of rich nations maintained special stores for their personnel and some of the merchandise found its way to other embassies as well. Social life among diplomats was fairly lively; invitations to banquets at embassies with a good cuisine were highly appreciated.[10]

Szekfü had been appointed to the Moscow post in November 1945 but it was only in March 1946 that he presented his credentials. Some of the delay was due to the fact that his first letter of appointment contained a number of mistakes, in outward form as well as spelling, and he feared that this might create resentment on the soviets' part. "It is well known," he wrote home, "that nowhere is there greater emphasis on observance of protocol formalities than precisely in the Soviet Union."[11]

In Washington the Hungarian representation had different problems. The first envoy, Aladár Szegedy-Maszák, was conscientious, hardworking, and loyal, but his political orientation was too far to the right. Mátyás Rákosi saw to it that Endre Sik, the cultural attaché and a reliable communist, kept a watchful eye on him. Sik addressed his letters directly to Rákosi. Acknowledging the sterling features of the minister's character, Sik nevertheless noticed that he much feared a shift to the left in Hungary and that the country would become "a member state [of the USSR] or something similar and collectivization, etcetera would follow." Szegedy-Maszák also feared that if Hungary got a peace treaty that did not contain territorial concessions, this would serve as grist for the mill of the reaction and that in turn would produce a backlash from the left.

Of the other two high officials at the legation Sik had little good to say. They were dishonest and dangerous people. "Both are careerists, both want to ride two horses at once and ingratiate themselves with both rightist and leftist circles."[12]

Another Rákosi confidant in Washington, Sándor Százas, complained that the legation was not receiving instructions from Budapest as to the political line to follow. This was all the more regrettable as the international situation had greatly worsened since the war. Two influential American publications, *LIFE* and *U.S. News & World Report*, had lately drawn parallels between the foreign policy of Nazi Germany and that of the USSR and were speculating about the possibility of World War III. While such talk was not to be taken seriously, it should not be lightly dismissed either. The latest Gallup Poll showed that 58 percent of the Americans interviewed ascribed goals of world domination to the USSR. Probably this was the reason why communists had lost so heavily in recent West European elections. The Americans hoped that the soviets would be restrained from an adventurist course by pressing manpower needs. The latest Five Year Plan did not permit the stationing of great numbers of able-bodied people outside the USSR. For instance, occupation forces in Hungary had been reduced by 60 percent (and this before the signing of the peace treaty).

Among soviet satellites Yugoslavia was viewed with least favor in the United States, mainly because of Josip Broz Tito's unswerving prosoviet policies and his rumored readiness to take Trieste by force if the peacemakers did not assign it to Yugoslavia. Rumania also had bad press because of its procrastination in carrying out free elections. Even Hungary's stock had fallen (the date of the report was June 1946) because its democracy was not developing in the hoped-for direction. "We are victims of the East-West struggle," Százas wrote. "The United States wanted a three-power review of Hungary's economic condition and a three-power effort to rehabilitate it, but the USSR did not cooperate. Yet without soviet concurrence we cannot expect any help; the United States had made it clear that it would not indirectly finance soviet reparations."[13]

While in the western capitals most Hungarian diplomats were still from the old school, in people's democracies they were generally members of one of the workers' parties. In the eyes of foreign ministry personnel in Budapest, there was a distinct, though entirely informal, pecking order among these socialist countries. The USSR naturally towered above all others and hence the selection of ambassador to Moscow was a particularly sensitive one. Gyula Szekfü was living evidence that at any rate in the first two years of the "new" Hungary the Smallholders controlled not only the foreign ministry but the foreign service as well. His past was unquestionably reactionary; at the same time he had such a protean character that a selective résumé could credibly present him as a progressive.

The second most respected people's democracy, until the summer of 1948, was Yugoslavia. Even the soviets held its "democracy from below" model as the one to follow. At the same time that model was anathema to the Smallholders, who were likely to be its first victims—thus in the first two years of the "new" Hungary Belgrade was a dubious post. That changed after the Kovács-Nagy affair reduced the Smallholders' Party (KGP) to a fragile shell of its previous commanding presence on the political scene. The new minister, was a communist and, in September 1947, he wrote: "Since the changes in the leadership of the Smallholders' Party and our reconstruction in a democratic direction, there has been a manifest increase in Yugoslav trust in Hungarian democracy. Lately, when they speak of friendly states, they never fail to mention Hungarian political leaders and the [Hungarian] press." He quoted the organ of the Yugoslav Communist Party, *Borba*: "Characteristic of our relations with Hungary after the war is the lively interest with which the people of Yugoslavia observe the efforts of Hungarian popular forces to democratize the country and contribute to the consolidation of peace in the Danube valley."[14]

That same month the envoy, Zoltán Szántó, reported on the enthusiasm with which Yugoslavs greeted the outcome of recent Hungarian elections (which, it will be remembered, were marred by irregularities). The coalition, wrote Yugoslav papers, had proved strong enough to settle accounts with the Horthy conspirators and save Hungary from new adventures into which the enemies of democracy, mercenary agents, and spies—Nagy and the others—had tried to plunge it. Moreover: "Election results show that Hungary is strong enough to resist designs of those western circles which even today openly support and revive the Fascist [system], in order . . . to lock the peoples of Europe into a prison of colonies."[15]

Where relations among communist parties (as distinct from intergovernmental relations) were concerned, the Czechoslovak party was probably most highly regarded in Hungary, even as popular sentiments were strongly hostile to the republic. The standing of men like Klement Gottwald, Zdenek Fierlinger, and Rudolf Slansky more than equaled that of a Rákosi, an Ernö Gerö, a József Révai or a László Rajk. All had come to full maturity in the Comintern in Moscow. Rákosi, as we have seen, was quite ready to subordinate the question of the ethnic Magyars, despite their victimization, to the demands of Communist Party solidarity. That he left literally hundreds of complaining letters from Hungarians in Slovakia unanswered was probably due to overwork, but it is nevertheless symptomatic. In his eyes possibly minority questions were an outgrowth of ruinous chauvinism.

With Poland contacts were spotty and inconsequential; it was not a neighbor and there were no outstanding questions to settle. The same applied to Bulgaria; the Hungarian attitude toward the parties in those lands could best be described as indifferent. As for Rumania, it stood (except for Albania) lowest in Hungarian esteem, not so much on the party as on the national level. Most Hungarians looked on their eastern neighbors as crude, unlettered, primitive mountain folk who had lived in historic Hungary for a thousand years without contributing to its culture and progress. The Rumanians reciprocated by branding Hungarians oppressors who had for centuries denied them the opportunity to develop their native gifts and impulses.

In this instance too communists tried to rise above hardened perceptions and ingrained antipathies; but it was an uphill battle and the crest of
the hill was, as it is today, out of sight. Countless studies emanated from the
political and cultural sections of both the Hungarian and Rumanian party
organizations, purporting to show that there was an abiding community of
interest between the two nations—that they, as one study put it, "fought
together against Turks and Habsburgs . . . [and] for social progress, in upris-
ings against the [reactionary] social order." According to the official party
line it fell to the working class in both nations to put Hungarian-Rumanian
relations on a new foundation. Now at last "the great victory of the Soviet
Union in World War II made it possible in both countries to lay the bases
for a more progressive and developed social order."[16]

Most of these studies remained buried in the party archives; even when
they did see the light of day, they were largely ignored by the populace.
It took far more than party rhetoric to kindle a brotherly feeling for Rumanians in Hungary or vice versa. A Hungarian Minister in Bucharest,
in one of his dispatches, complained about the condescension with which
the Hungarian public and media treated the Rumanian nation: they practically ignored developments in that country and the press hardly ever
published an item about political or cultural events there. By contrast
the Rumanians evinced the keenest interest in Hungarian affairs and frequently commented on them.[17]

The insecurity of Rumanians vis-à-vis their western neighbor was evident in so many large and small ways that it could easily form the subject of
a monograph. Their politics, often by their own admission, lacked character
and initiative. Another legation report, by the same Minister, is instructive.
Shortly after Tito's expulsion from the Cominform a Rumanian journalist
submitted an article about Hungary to a newspaper. The editor rejected it,
saying that he had to be careful with regard to all material relating to Hungary. "It was probably not one man's caution," the Minister wrote, "but a
new manifestation of Rumanian attitude. Since the resolution of the
Cominform, Rumanian officialdom has withdrawn into extreme reserve
toward all people's democracies; it intensifies its contacts only toward the
Soviet Union." (Curiously this report was addressed to László Rajk at a
time when he was already under arrest for Titoist activities.)[18]

The Minister, Jenö Széll, enjoyed an unusually long tenure in Bucharest,
perhaps because he was able to balance his patriotism with an understanding for the people to which he was accredited and avoided partisan prejudices. His replacement was a colorful and plucky young man, Iván Kállo,
whose reports rank among the most entertaining, if not the most substan-

tive. A former lathe operator, he took a communist's pride in his humble beginnings. His reports were so frequent and so lengthy that it was hard to believe he found time for anything else than writing them, yet he assiduously attended to the business of the legation as well as to numerous social occasions.

He made it his business to pay a courtesy visit to every legation in town. The British Minister, when he learned that Kállo was a communist, gently teased him: "Of course, now you have to be careful about what or how much you tell me. . . . I mention this because I count as an imperialist and a reactionary who one must be wary of." He added that this was probably the reason why he so seldom had visits from diplomats representing people's democracies.[19]

The head of the Rumanian Planning Office, when Kállo visited him, gave evidence of the high regard in which his nation held its western neighbor. He commented on how enviably skilled Hungarians were in their fields and how eager to learn; he said he wished Rumanians had been like that. But, he admitted, Rumania was backward and suffered from a dearth of learned tradesmen. Also, in Hungary the currency had been stabilized in 1946 and a three-year plan had been launched in 1947. In Rumania both took a year longer and the plan stretched to five years.[20]

The finance minister, when Kállo visited him, voiced similar complaints. Unlike in Hungary, he said, in Rumania collectivization was very difficult to carry out; a general survey of who owned how much land had been conducted three times and each time yielded different figures. The peasants were sly; they bought and sold each other's land, eluded surveys, and evaded taxes.

Kállo's visit to the American envoy, Rudolf Schoenfeld (brother of Arthur who had been Minister in Budapest), proved to be an unpleasant one. Schoenfeld was even more sardonic than his British colleague had been. When Kállo told him that he had once been a manual worker and was now a diplomat, Schoenfeld acidly remarked, "Well, related fields." Kállo reported home: "Entire discussion, in German . . . lasted eight to ten minutes."[21]

He had another ideological clash with the Belgian Minister, Alfred Herman, who came to visit his legation. Herman related with pride how prosperous life was in Belgium, how much people earned, and how well they dressed. Kállo pointedly inquired how life was in the (Belgian) Congo. Herman either didn't catch the irony or ignored it. The Congo was an important part of Belgium, he replied. Belgian life would be unimaginable without it. He went on to say that on that enormous territory lived about 1.5 million people, of whom 800,000 were "Negroes," most of them unsuited for work because they were stunted in growth. The reason for that, Herman explained, was that in the age of slavery the healthy and young had been taken away; only the unfit were left behind. But whites and Negroes now freely intermingled and the half-breeds were robust. Kállo asked him, if this was the case, why was intermarriage not sanctioned? Because, the Minister replied, there was too much venereal disease among the Negroes. The Belgian government spent large amounts trying to eradicate it. Kállo asked him whether all the profits from oil and rubber could not be used for that purpose. Herman left that question unanswered.[22]

The Minister from the German Democratic Republic, Ionni Löhr, proved more than a match for the free-speaking Kállo. A fierce communist, he had spent many years in prewar Rumania, most of them in prison. Many men now in leading positions had been incarcerated with him. He told Kállo that in the Democratic Republic, which made up 20 percent of prewar Germany, people were enormously grateful to the Soviet Union. He spoke with utter contempt of the "Adenauer clique" now ruling the western part of Germany. He explained Konrad Adenauer's "filthy treason" with the fact that his wife and "Mrs. Klay" (sic) (wife of General Lucius Clay) were sisters or cousins. But even in that part of Germany, Löhr contended, people increasingly allied themselves with the Communist Party.[23]

When Kállo returned Löhr's visit, the latter treated him to another outlandish tale. There was an active effort in West Germany, he said, to recruit people into the French Foreign Legion to relieve the terrible unemployment. Magazines carried open advertisements depicting the luxurious lifestyle of young people serving in "Indonesia" (Indochina). There were pictures of a 16-year-old girl in a sumptuous apartment, with suggestions that others signing up would find the same opulence. Germans going there were promised cushy administrative positions and were assured that they needed to have no fear of reprisals from the natives as troublemakers were simply shot. Löhr commented: "If I didn't feel sorry for those young people I would say that they got what they deserved. Many are seduced by American cigarettes, boys and girls smoke freely, many girls are inveigled with promises of a rich life in America." There was unfortunately a great lack of reliable party cadre in the western part of Germany; Löhr hoped some good communists would return from the prisoner-of-war camps in the Soviet Union, where they had received the necessary education.[24]

The dearth of trained and disciplined party men was a problem everywhere in the soviet orbit. To be sure, the requirements for membership were exacting—peasant or proletarian birth, good intelligence, a reasonably thorough Marxist education, and unbending discipline—and such individuals were not easy to find. A disproportionately large number of party members were Jewish. In their case the requirement of a humble birth was generally dispensed with. Many looked on communism as the system most resolutely opposed to the regimes that victimized them. Yet just now another cause attracted many Jewish loyalists: the fate of the State of Israel. The USSR, which had been the first nation to grant recognition to the Jewish state, soon realized the potential for dual loyalty and began a campaign against Zionism and "cosmopolitanism." Kállo remarked in one of his reports that the growing volume of emigration from Rumania to Israel produced acute problems, especially in intellectual fields. Not only did it fan the flames of anti-Semitism but also many of those seeking to emigrate were party members and their places were difficult to fill. The entire personnel of the State Publishing Office had signed up for emigration. Also, many engineers, badly needed in heavy industry, as well as technicians, planned to move to Israel. In the past such people were immediately dismissed from their jobs—but the state could no longer afford to do that. Government officials were trying to prevail on would-be emigrants to stay by keeping them on their jobs until the last minute. The press was carrying articles about the wretched conditions in Israel and it spoke of Zionists as handmaidens of the United States.

Kállo's reports at times verged on the comical. He sent a smugly malicious account of a botched reception at the Czechoslovak legation; the orchestra played so long that two of the most important personages present, the foreign minister and the Russian envoy, dozed off. When strawberries were served they were so sour that they had to be sprinkled with confectionery sugar; the foreign minister Anna Pauker ate one and began to retch; the sugar turned out to be salt.[25]

The diplomatic reports from abroad routinely drew comments from a foreign ministry official in the competent bureau and more often than not the comments were critical. Kállo did not take to this practice kindly. His cadre rating was apparently high enough to put fear of authority aside. In one of his rebuttals he spoke out against the "old, stodgy diplomatic usage," both at legations and at the ministry. There was no reason why comrades should communicate in this fashion, he wrote. "I wish to call attention of the ministry to the fact," he once wrote, "that it too seems to feel bound by these ossified habits. I am being censured over bagatelle issues. Such pedantry is anachronistic, uncomradely and will not in the least influence my work at the legation."[26]

Meanwhile at the Washington legation things were going from bad to worse. The damage caused by a chain reaction of resignations after Ferenc Nagy's ouster was never repaired. No sooner did a new contingent of diplomats and service personnel settle in than a part of the previous one defected;

also many of the new appointees had to be dismissed or demoted for incompetence. The turnover was embarrassing. Rusztem Vámbéry, who succeeded Szegedy-Maszák (after a brief tenure by Pál Marik as chargé), was a stately superannuated gentleman who, according to Endre Sik, spent his time making official visits; after several months he went on sick leave and subsequently defected. Another high official was found to intrigue with the State Department and had to be recalled. Still another was found unsuited for legation work and was transferred to a consulate. Endre Sik, who followed Vámbéry as Minister, found himself and his staff unable to submit a summary report for 1948 because, as he put it, the work of the legation was "marked by paralysis." In addition to the constant personnel changes there had been a terrorist attack against the premises that disrupted normal work. Somewhat later Budapest sent out a woman who, while politically reliable, proved so difficult that she totally disorganized the work of the legation.[27]

Endre Sik's communist principles finally overcame his administrative concerns. He looked on the constant turnover as a purge and wrote to Rákosi that the more thorough it was the better. But he could not gloss over the fact that the American representation of the Hungarian People's Republic was practically defunct. As he reported, there was only one employee who could speak a broken English, and he was about to be recalled. "It has come to this," he wrote, "that the State Department or the FBI can provoke us or offend our prestige and we can't fight back." At the same time dissidents like Ferenc Nagy had free access to American authorities. At the consulates matters were, if anything, even worse than at the legation. Everybody wanted to be the boss, everybody was overweening and overly sensitive. The new personnel looked on the old as politically unreliable intellectuals alien to the working class and thus often withheld crucial information from it.

(Rákosi forwarded the letter to Andor Berei with a note saying, "Urgently request suggestions." Berei recommended a tried communist, Imre Horváth, for the Washington post and another old party member, Ida Gyulai, as a counselor. When Sik was advised he gratefully agreed.[28])

At last the legation struck a balance between political reliability and administrative competence but, in the years that followed, the former remained always the principal consideration. The annual summary reports amply demonstrated that. Their concepts and phraseology were so hackneyed and predictable that it was unclear what purpose those often lengthy documents served. They were without exception for internal use and were preaching to the converted. At times they contained some valid observations, for instance that the United States after the Second World War sought to squeeze other capitalist states out of their competitive positions and that the main obstacle to the American goal of global dominance was the USSR. The brunt of the arguments however was always the same, that the United States was hostile to true democracy and had opposed the new Hungary from the very beginning. As one example a report cited the fact that some 400 war criminals whose extradition Hungary had demanded were not handed over, often with transparent excuses. "This shows," the report averred, "that already in 1945 the United States had far-reaching plans with fascistic emigrants, hoping to use them to restore the old order in Hungary."

Another paragraph of the report charged that as early as 1945 the U.S. legation in Budapest had began recruiting its Hungarian employees for espionage. "On various pretexts they began to interfere with our domestic affairs, going so far as to organize and support conspiracies."[29]

A 1949 dispatch from the Paris legation reveals that during the struggle between József Mindszenty and the Hungarian People's Republic an alternate Catholic Church came into being in the country that established a comfortable modus vivendi with the government. Two clergymen of this collaborationist church visited France and had a lengthy talk with a French priest, a Father Boegner. When Boegner asked them if there was religious freedom in Hungary, he received an emphatic positive reply. He was then treated to a historic retrospective in which the visiting priests reminded him that in France the bourgeois revolution (preparatory to the socialist revolution to be followed in the end by the communist) had long taken place; not so in Hungary, which was now moving toward socialism in a straight line. It was making its peace with the Catholic Church and Father Boegner could rest assured that that church had chosen a path laid down by God. Whoever tried to divert it from that path sinned against God. That had been Mindszenty's sin. "Our own peaceful and constructive cooperation with the government is proof that the road we had chosen is blessed by God," the clergymen concluded.[30]

Many legation reports, formal and informal, bristled with personal animosities. The Moscow mission was notorious in that respect. One report, from the pen of a woman employee, had as its target the ambassador himself. András Szobek, as we have noted, began his political career in the ephemeral Hungarian soviet republic of 1919, later joined the Social Democratic Party (SDP) and served as a provincial secretary for some twenty years. After World War II he switched to the politically more useful Hungarian Communist Party (MKP) and held several posts before being appointed ambassador in Moscow. He was apparently a jaundiced, ill-humored man.

According to the woman's report, when Rákosi and his wife came to Moscow to help celebrate Stalin's seventieth birthday, they stayed in

Szobek's apartment. Before the festivities the woman was asked to help Mrs. Rákosi dress. She was honored to be able to assist the party chief's wife but Szobek said to her later: "What does Rákosi think, I keep servants for his use who will tiptoe around him and be of his service?" He spoke of the soviet people with contempt, calling them lazy and indifferent, and his scorn extended to the embassy employees and civil servants. When a soviet diplomat was scheduled to leave for Budapest to take up a post there, Szobek and another embassy employee went to the railroad station to see him off. The diplomat was late. The employee accompanying Szobek remarked that he was probably checking his baggage. Szobek replied with asperity, "Do you think these [people] have baggage? They have nothing. But just wait, they will have plenty when they come back." Of another soviet official Szobek remarked, "A Jew, a filthy Jew, like his whole race."[31]

The world of diplomats, like any other outwardly dignified establishment, naturally has its seamy side. Backbiting, personal calumny, sexual misadventures, and abuse of privilege are common. So is the use of diplomatic immunity for intelligence work. In communist missions to all this was added the ideological straitjacket, which skewed diplomatic reports and smothered objectivity. In fact any opinion, any presentation of facts, that did not conform to the current party line was regarded as a sign of immaturity, a lack of erudition. When a diplomat at the Washington legation submitted a study on how, according to reliable intelligence, the United States was stockpiling strategic materials in order to be independent of outside sources in case of war, the reader of his study in Budapest upbraided him for "adopting a certain theory [formulated by] a bourgeois economist." As to the "true" motive for the stockpiling: "Is the problem really that the state lacks sufficient raw materials? No. It is rather that the government wants to eliminate the effects of objectively functioning economic laws [inherent] in modern capitalism."[32]

An undated and unsigned prescript from the foreign ministry set down guidelines for communist foreign policy. It started by recommending an ideologically serviceable vocabulary. "In outlining our program the greatest care must be taken in the use of every word and phrase, because even the most seasoned communist personages are inclined to employ Fascist [i.e. traditional] terminology." In general, practitioners of the "new" diplomacy were cautioned to avoid using words that had no concrete meaning behind them, such as "democratic, "social," or "self-governing." But the study at once involved itself in its own obscurities. "We want the dictatorship of the proletariat. . . . [This] term however gains a new meaning in the dialectic interpretation of the experience of the past decades. In the apparently 'parliamentary,' 'liberal,' 'democratic' order so-called democracy was nothing else than the dictatorship of the bourgeoisie. In contradistinction, Lenin emphasized the dictatorship of the proletariat . . . which is in fact the only democracy."[33]

Not only officialdom but also the public had to learn to view global events through the terms of this new lexicon. Seemingly ordinary words acquired a "true" meaning when understood correctly. A separate study on the significance of the Rajk affair analyzed the meaning of the term Trotskyism, in its former as well as its current context. The study reminded its readers that Stalin had commented on the Moscow trials of 1938, saying that "today's Trotskyism differs substantially from the revolutionary one." The distinction valid in 1938 was still applicable. "Today's Trotskyism is not merely a mistaken or malignant policy direction aimed against the party but a spy apparatus of imperialism."

That Leon Trotsky was an imperialist agent had of course never been proven, and even less that Tito or Rajk was. No matter. Trotskyism had become a generic term covering all political evil.

Rákosi after the Rajk trial devoted much verbiage to the lessons that the case provided for future generations of communists. In a speech of September 30, 1949 he confessed to his audience that it had taken him many sleepless nights to work out the plan by which the Rajk gang could be liquidated. Now that the case was closed it served as an alert to communist parties within the soviet orbit and in capitalist states not to fail to search out such conspiracies. Moreover (here was a warning to all those who thought they could now relax their guard), "it would be in vain to believe that by rolling up the Rajk conspiracy we have totally liquidated . . . the enemy's organizations. We know that it has taken the enemy decades to build up its spy network. It is unlikely that what it took years to erect we can demolish in a few months."[34]

The Rajk case was indeed only the beginning. Not even the darkest period of "reactionary" terror, in the 1920s and 1930s, witnessed as many hangings and imprisonments of devout communists as did the early 1950s. The revolution was devouring its own children.

Chapter 26

Korea and the View from the Peace Camp

Although diplomatic reports from Moscow continued to exult over the growing strength and prestige of the USSR in the world arena, 1949 had proved a dismal year for soviet foreign policy. Of the two events in the credit column only one truly enhanced its power position: the detonation of the first soviet atomic bomb, over Siberia. The other, Mao Tse-tung's victory in China, was in line with Stalin's scenario for world conquest, which predicted that Marxism would march through the colonial areas first and that the sovietization of China would be the first great leap on that road. However many signs indicate that Stalin was made uneasy about the appearance of a new communist superpower—one that, after its consolidation, would be able to lead and control the communist movement in Asia.

In Europe communism had been contained. What had started as a West European defense system had grown into the North Atlantic Treaty Organization (NATO), in which the United States was the dominant power. The adherence of Norway and Iceland brought the vital northern waterways under NATO's purview and Italian membership, although at first less than welcome to Americans, ensured a powerful naval presence in the Mediterranean. Soviet fulminations only served to make the alliance stronger. The attempt to squeeze the western powers out of Berlin had failed. By the time

Stalin lifted the blockade Western Germany had become a separate and partially independent state. It was only a matter of time before this new Germany would be incorporated into the western defense system. Stalin's worst fears would then be realized. Within the soviet orbit Josip Broz Tito persisted in his defiance of the USSR and received increasing aid from the west.

Hard-pressed strategically, the soviets began to stress the theme of peace with a heavy-handed but unrelenting propaganda offensive. Here at last was a theme with which the war-wary and undernourished masses could identify, one they were ready to accept at its face value. In its most comprehensive interpretation it could even mean that for the sake of global peace the soviets were ready to abandon their program of world revolution. During 1950 there was barely a communist-sponsored demonstration, commemoration, or anniversary celebration that did not resound with some variation of the theme of peace. An international peace conference in Stockholm was attended overwhelmingly by members of progressive leftist organizations.

Mátyás Rákosi's secret papers contain a summary of the Hungarian Workers' Party's (MDP) efforts to promote the movement. By direction of the politburo every foreign policy article in *Szabad Nép* had to be linked in some way to the theme of peace. The paper had started a special column, "Struggle For Peace"; other press organs adopted the practice. Meanwhile a slew of articles contrasted everyday life in the United States, permeated with militaristic thinking, with that in the Soviet Union, inspired by the common desire for friendly relations among nations. In its March 5 issue *Szabad Nép* carried an article entitled "Moscow-Pittsburgh" devoted to this subject. In Pittsburgh the working classes lived in squalor; in Moscow the prices of consumer goods were being constantly lowered. In another article the paper reported on the struggle of colonies for liberation. Ample space was also reserved for the denunciation of Tito and his renegade party. Tito stood in the pay of warmongers; during Stalin's birthday celebrations the main slogan was: "Stalin Means Peace."[1]

April 4, 1950 was the fifth anniversary of Hungary's "liberation." Long in advance the MDP's local organizations staged demonstrations hailing the noble endeavors of the peace camp, against which the warlike imperialists were helpless. An MDP memorandum nevertheless noted that this outpouring of emotion carried a danger: it led many to believe that the forces of aggression had been destroyed and that henceforth the preservation of peace was the business of the government, which could achieve it by strengthening the armed forces. This was a mistaken notion. The struggle had to involve everybody. Peace was the responsibility not of the armed forces but of progressive mankind. The MDP propaganda section began circulating petitions (Rákosi's signature was the first one appended to it) pleading for peace. Posters depicted a child in a Pioneer uniform holding out a sheet of paper to his mother. The legend read, "Mommy please sign!"

Mommy might have signed but, as the propaganda section noted, the reaction—still alive, still truculent—and its agents, the Catholic Church, the kulaks, Horthyist military officers, and gendarmes, not only refused to sign the petition but often openly condemned it.[2]

A security police (ÁVO) report of May 4 pointed out that the churches had tried to keep people away from the April 4 celebrations. They abetted workers to celebrate Easter instead. According to the report one Franciscan "general" had proclaimed: "At services we have to pray for the poor Hungarian nation and for the coming of its true liberation." The vicar general of Esztergom (the onetime seat of Cardinal József Mindszenty) had urged worshippers to make the April 4 services different from ordinary weekday masses. "We should pray for the Fatherland; liberation by the soviet army should not be mentioned."[3]

Later in the year Erik Molnár, now minister of justice, sponsored a bill in parliament entitled "For the Defense of Peace." It began: "The National Assembly of Hungary, conscious that war is the source of immeasurable suffering for mankind and the greatest obstacle to social progress . . ." and concluded that "the removal of war danger is, regardless of ideological convictions, the common interest of all mankind, and ensuring peace requires the cooperation of the world's peace-loving peoples. . . . The Hungarian People's Republic will strike with full vigor of the law against all those who imperil peace among nations." The bill provided that anyone who, orally or in writing, through the press, the radio, film, or any other media agitated for war or promoted war propaganda, was subject to imprisonment of up to 15 years and the confiscation of his wealth.

The peace propaganda was put to a severe test by the outbreak of the war in Korea. That the war was launched by the People's Republic in the north was common knowledge. It took a dialectic sleight of hand on the part of communist propaganda to contend the opposite. The starting point was that there was only one Korea, hence an invasion of one by the other was a logical absurdity. Since imperialist forces had usurped the southern portion of the state and set up a puppet regime that scotched democratic aspirations, the so-called invasion was a spontaneous uprising of the people against foreign rule. It was thus not war but an internal struggle for liberation.

The Hungarian government, now entirely dominated by communists, adopted this argumentation wholesale. The true aggressor, it held, was the

United States, aiming to restore the reactionary governments of the past. Endre Sik, now director of the political division of the foreign ministry, composed a circular for missions abroad. The central theme was that "the imperialists had passed from war preparations to outright aggression and they use the United Nations to legitimize their aggression." Diplomats should point out that in American satellite countries, because of the sharp economic downturn and the growing resistance of the working masses, the political crisis was steadily deepening.[4] (Unfortunately by now the theme of an impending capitalist crisis was wearing thin, especially as there was no evident sign of it, although it could be argued that the Korean War came at a time when recession threatened the American economy and was helpful in averting it.)

The British Labour Government, an American satellite, was going to great lengths to break down the peace movement—this was another contention of Endre Sik's circular. The working class resisted the effort by giving the currently conducted wage negotiations a political character. Sik requested that the legation in London report on this latter development and on what part the British Communist Party was playing in it, as well as on how working people reacted to the illegal methods by which the United States was using its voting machine in the United Nations.[5]

In his person and in his outlook Endre Sik epitomized the conundrum in which men of breeding and conscience found themselves as they tried to serve the "new" Hungary. A man of wide culture, he was a former Piarist seminary student, a jurist, an author, and a historian. He had turned to communism as a prisoner of war in Siberia in 1919-1920, but he retained a balanced outlook. He combined in his thinking the ideals of socialist progress with a tolerant world view that is the hallmark of a cultured man. But totalitarian ideology, especially of Stalin's brand, did not allow such synthesis of values. Sik, like others who wished to remain active in political life, had to choose; by 1950 the last vestiges of political pluralism had vanished.

Diplomacy had shrunk to obeisant subservience to the world policy of the Soviet Union on the one hand and spiteful skirmishing with the western powers on the other. In the 1950 series of the *Foreign Relations of the United States* (FRUS) there is only a brief chapter on relations with Hungary and it contains almost exclusively documents relating to efforts to secure Robert Vogeler's release. Contacts over other issues—political, economic, or cultural—had come to a complete halt.

At the end of 1949 a working group of experts in the Southeast European Office of the State Department prepared a policy statement relating to Hungary; at one point it stated: "Although we maintain diplomatic relations with the present government of Hungary, we will periodically review the advisability of this relationship, keeping in mind the diminishing effectiveness of U.S. representation in Budapest due to hostility and obstruction on the part of the Communist regime and the gradual drying up of sources of intelligence." The economic portion of the report noted that: "American-owned properties and interests had been lost to their owners without compensation through outright expropriation . . . transfers to the USSR as 'German assets,' sudden imposition of excessive taxes intended to induce bankruptcy, false charges of 'economic sabotage' or through simple seizure of premises."[6]

When in December 1950 Nathaniel Davis submitted his impressions of the year just passed he could only register a further worsening of relations, and offer a nostalgic retrospective to times when things were different.

> In the period immediately following the cessation of hostilities . . . contacts between American representatives and Hungarians, both in and out of Government, were widespread and friendly. It was not long, however, before pressures from the Hungarian side, implemented particularly by the Security Police, rendered unofficial contacts increasingly difficult to maintain while at the same time officials of the Government gradually withdrew from even the most casual social contacts with Western, and particularly American and British, representatives, while official exchanges on matters of business were gradually restricted to a few officials of the Foreign Office.

Davis noted that after his arrival in Budapest he had tried to obtain appointments with members of the cabinet. He was assured that such appointments would be made. In the end Rákosi was the only one to grant him an interview and then "only after more than 2 months of insistence and persistence on my part in connection with the Vogeler case." Davis continued:

> The disinclination or fear on the part of Hungarian officials to be seen . . . in my company was dramatically illustrated within a few weeks after my arrival. At a cocktail party offered by the Swedish Minister . . . I was introduced by a Foreign Office acquaintance to the Under Secretary and the real power in the Ministry of Finance. As soon as this gentleman learned my identity he not only withdrew his proffered hand but withdrew from my presence, went straight to the lobby, retrieved his hat and coat and departed.[7]

When the Hungarian legation in London reported on relations with His Majesty's Government (HMG) in the year 1950, it could only recite the many areas of friction, how the arrest and trial of Edgar Sanders had brought the commercial and financial talks to an end, how the British (and American, Australian, and New Zealand) charges of Hungary's violation of the peace treaty further aggravated relations—and how, on the reverse side, the legation had broadened and intensified its contacts with progressive circles in Britain, with the peace movement, and with persons of an anti-American bent. The White Book issued by the Budapest government on the Sanders affair had attracted so much attention that the first printing could not satisfy the demand. Another book, entitled *Tito's Plot Against Europe*, appeared at the end of 1949 and also had to be reprinted.[8]

In Washington, Endre Sik's successor, Imre Horváth, had inherited the task of rebuilding the legation from its depleted and demoralized state. (He could forget about the consulates for now: they had been closed down in the wake of the Vogeler affair.) He had arrived at his post in October 1949 in such a hurry that he could not bring his family along. (He returned to Budapest some time later to arrange for his wife and children to join him.) Dean Acheson, when Horváth came to visit him, impressed on the envoy the importance the State Department attached to Vogeler's release. Horváth promised to transmit the message to his government and added that there was little he himself could do in the matter.[9]

The legation continued to be plagued by incompetence, but no longer by ideological dissonance. The new staff had strict guidelines for the evaluation of any event. According to these there were no truly free states left outside the "peace camp." The policies of West Europe and of the colonial areas were dictated by the capitalistic ruling class in the United States. The Korean War had produced a watershed because of the refusal of a number of American allies to participate in it. Before the war NATO had operated on the American-imposed principle that while military preparations were important, they should not be allowed to retard economic recovery. The new policy was that military readiness had to be achieved at any price. (This argument, which suggested that the new policy resulted from an unexpected turn of events, contradicted the assertion that the United States had premeditated the war in Korea, but that did not seem to trouble the authors of the new guidelines.) The American press, so the directive continued, repeatedly complained that the "Marshall states" did not share the burden of defending Europe. Truman had instructed his ambassadors in those states to demand an increase in military strength. Within a short time a force of 56 divisions had to be set up, of which 20 were to be armored. Originally the changeover from a peace to a war economy had been planned for 1956, but Washington now insisted that it be completed by 1952.

There were also reports that the United States planned to integrate Yugoslavia into the European war-bloc and would propose this move at a soon-to-be-held western foreign ministers' meeting in London. But differences of opinion on this question were so great that it had for now been removed from the agenda. In actual fact, while the American government had full faith in Tito, the Joint Chiefs of Staff doubted the loyalty of the Yugoslav army, as it had not been purged of its prosoviet elements.[10]

A subsequent report characterized the war in Korea as a purely capitalistic undertaking. According to the report political moderates in the United States had expected that when the "imperialist forces greatly superior in numbers" reached the 38th parallel, the aggression would stop, especially because it carried the risk of a world war. But "MacArthur's military clique," as well as war industry interests, opted for continued aggression. The appointment of General George Marshall as secretary of defense was significant because the general had close contacts with the Morgan financial group, which in turn had strong European ties. This meant that while the general policy of aggression would not cease, its brunt would shift to Europe. Rumor had it that "the infamous pro-Nazi commander-in-chief of American forces in Europe, Gen. McCloy, would in due course replace George Marshall as secretary of defense."

In summary, according to the report, rearmament had become the foundation of American economy. This was evident from the stock market reaction to rumors of peace late in September; there was a sudden fall in prices. But when the possibility of a broader war with China emerged, prices rose sharply. The feverish armaments program had for the time being put off the long-overdue economic crisis, but it also spelled dangers to capitalism: unchecked inflation that could rock the country's economic foundations; and the loss of recently acquired markets that could not be supplied as war industry exhausted the country's productive capacity.[11]

The Great Depression and international developments in the 1930s had left an indelible mark on communist thinking; the idea arose that capitalism, before it collapsed because of vanishing markets and the impoverishment of the masses, would pass through another economic and political phase: fascism. This new dogmatic element explained much that had happened in capitalist states, especially the United States, after World War II. Hungarian diplomats often commented in their reports about emerging fascist features in these states; they pointed however to the difference in the advent of fascism in the United States from the pattern in Italy and Germany. "The measure of popular resistance [in the United States]," noted one report, "will be determined by the strengthening of the peace camp which will decide whether fascism will fully triumph in the United States, and if yes, in which variety. The difference is that here [fascism] is not the product of a new party, a new capitalistic group, or even a new philosophy. It does not openly oppose bourgeois democracy and its constitution. There is no 'March on Rome,' no 'Machtergreifung' or even the pretense of a revolution. There is rather a gradual smothering of popular democracy."

Leading circles in America, the report observed, had bred a "workers' aristocracy." Most skilled workers had benefited from the surplus value produced by capital and their class consciousness had in consequence been eroded. The Communist Party was weak. The point at which a revolution was inevitable had not yet arrived. But the growing strength and prestige of the USSR and the international workers' movement, as well as the developing crisis in the capitalist world, ensured that old measures were no longer sufficient to keep the exploited masses in check. It was not enough to throw morsels of surplus profits to the discontented, or to heap calumny on the Soviet Union. The Korean War represented a step forward on the road to fascism. It provided the pretext for rearmament and excused a further lowering of the living standard. The result was a growing resentment on the part of the working class; but there was also an increase in chauvinism and war hysteria. Capital did not dare to challenge the workers' movement. It operated with superficially constitutional means. In this respect the Anglo-Saxon tradition served the purposes of capitalism well. The bourgeoisie achieved its success not by destroying the feudal system but by exploiting it.

However world affairs were not static. The peace camp was growing in strength; this led to shrinking markets and dwindling profits. Whether fascism would come to America depended on the degree of workers' resistance; the main question was however whether it would come by legal means or through violence and imposition.[12]

Such was the view from the Washington legation. Meanwhile the Moscow mission operated under even stricter guidelines. Budapest demanded that every report on the political situation include: (1) reference to the USSR as the leading force in the struggle against imperialism; (2) the official soviet position on all important international questions; (3) the support given by the Soviet Union to nations building socialism; 4) soviet participation in international organizations. Reports on the internal politics within the Soviet Union had to stress the role of the Bolshevik Party in major legislative enactments.

In line with these directions the report for 1950 stated that the year had passed in the struggle for peace. This was an imperative feature of soviet tradition. Lenin and Stalin had repeatedly pointed out that cooperation with capitalist states was not only possible but necessary. (The author of the directive should have read Stalin's February 19, 1946 speech that asserted the exact opposite.) The USSR strove for a lasting democratic peace. In its efforts the USSR continued to expose the aggressive plans of the warmongers and attempts by imperialists, especially the United States, to subjugate other states and create the conditions for a new war.

The success of the struggle for peace was made possible by the great economic strength of the USSR. In the course of the year the Stalinist Five-Year Plan had been completed. Industrial production grew 70 percent over the last peace year. The workforce had increased by two million since 1940. Of special note was the success of Stalin's nature-altering program. A number of new power stations had been constructed, canals and hydroelectric plants had been built. "These achievements of the Stalin era are without parallel individually; taken together they represent an aggregate such as the technological world resting on capitalist foundations did not and cannot imagine."

In foreign policy the Soviet Union was consistently condemning American aggression against North Korea and China. It also protested against the rearmament of Japan. It tried to extricate the Austrian peace treaty from the cul de sac into which it had been brought by western imperialists. In international organizations the Soviet Union vigorously protected people's democracies.[13]

One may wonder what insight the foreign ministry in Budapest gained from reports of this kind, reports whose format and content it had itself prescribed and which, even without the prescription, would have been a mindless glorification of the USSR and the soviet system. The purpose was not to present an objective analysis of soviet intentions and policies but to adhere to a manner of reportage that, although hollow and insubstantial, could not be censured on ideological grounds. Still, the foreign ministry at home always found faults in those reports. They were not political enough. Or thorough enough. Or perceptive enough. The only way the legation officer could anticipate such criticism was by raising the decibels of his praises and enriching them with Marxist quotations culled from Lenin's or Stalin's writings.

The introduction of such a standardized phraseology was one way for Moscow to keep a close watch over satellite diplomacy. It mandated such a copious flow of praise of soviet purposes as the only ones worthy of a democratic nation that little room was left for private or national initiative. The supervision was necessary because foreign policy tends to elude control; it cannot be planned and directed in the same way domestic policy can. The Tito affair showed that ideological orthodoxy did not ensure proper behavior; the slippage in the behavior of the Yugoslav party put the unity of the entire peace camp in danger. To prevent another defection, the foreign policy of the satellites had to be made a mere appendage of soviet global policy. Its main function was not to foster constructive relations with other nations but to demonstrate the intense attraction that the Soviet Union and its ideals had on nations that had shaken off the burden of the past and entered the road to socialism.

Yet propaganda goes only so far and cannot conceal decay. The hysterical tenor of the anti-Tito campaign was a symptom of the terminal phase of Stalinism. All that had been dynamic and creative in Stalin's work was exhausted; the Five Year Plans with their wearisome statistics and self-congratulatory celebrations no longer fired a jaded public that earned only the most meager rewards for its mighty exertions. The crisis of capitalism had become a mere empty slogan. The Council For Mutual Economic Aid (COMECON) produced more friction than cooperation. Recent Stalinist initiatives—the Berlin blockade, the denunciation of the Marshall Plan, the Korean War—had all proved costly fiascos. After 20 years of a personality cult that reached its apogee in the overblown birthday celebrations of the Great Leader, there followed an almost palpable anticlimax, as if the supply of hyperboles had been used up and only soulless repetition were possible. When Stalin was proclaimed by some worshippers as the world's greatest scientist, even devout Stalinists recoiled in disbelief.

Still, Stalinism had to be sustained because there was no replacement, even though the fight against Titoism was now its only focus and terror its only means. Nothing better revealed the decay than that the sole unchallenged leader of the communist camp, the proclaimed liberator of the laboring masses the world over, turned with single-minded fury on the head of a medium-sized state and lent his immense prestige to a propaganda campaign that relegated all other causes to a secondary status, as if under the heading of Titoism the other adverse forces—capitalism, imperialism, exploitation, treachery, and conspiracy against the working class—were subsumed.

That the foreign ministry in Budapest, and those in other satellite capitals, found the time to turn out one well-documented "study" after another to demonstrate the crimes of Tito and his "clique" showed that there was little else useful to occupy the personnel. All the normal activities, such as the collection of data, policy planning, report analysis, and the preparation of position papers, were neglected, and it was just as well because they were not needed anyway. There was but one foreign policy apparatus in the soviet camp; the foreign ministries in the satellites served as subordinate bureaux with not even a vestige of autonomy.

Conclusion

The central question emerging from this study is whether Hungary in the wake of World War II was slated for a special place and special role in the East European state system—a place and role that eventually was denied it because of adverse international developments—or whether from the beginning it was intended to share the fate of other soviet-bloc nations, only in its case the process was delayed because of peculiar local conditions. Alternately, did Stalin perchance intend to leave the fate of Hungary to the free play of political forces as a novel experiment in an otherwise uniform pattern of soviet imperialism? Still another possible version is the one Mátyás Rákosi confided to his comrades, that Stalin planned to synchronize the sovietization of Hungary with the final crisis of western capitalism. The arguments in favor of each theory are so finely balanced that no conclusive inference is possible.

In favor of the special-status thesis we have only one direct evidence and that is thirdhand: Rusztem Vámbéry's intimation to a State Department official that, according to soviet envoy Georgy Pushkin, Moscow intended Hungary to become a bridge between east and west and hence would allow it a measure of independence that would make the country an attractive prospect for western investment. It is unlikely that Vámbéry invented this version, mainly because its second part, that the soviets wanted a prosperous Hungary only better to be able to exploit it, vitiated the value of the first. The same reasoning applies to Pushkin's communication. The only uncertain element is whether Pushkin was privy to the Kremlin's designs or whether he gave Vámbéry his own opinion.

The proof for the opposite theory is readily at hand: within less than five years of its liberation, Hungary was a sovietized state solidly built into the eastern bloc under Moscow's dictation. The slower pace and halting pattern in the case of Hungary can be explained by tactical considerations that however in no way affected the strategic plan.

In support of this version speak the numerous reports by western ambassadors, all pointing to the suppression of opposition voices and the essential monopoly of political power by the Communist Party (although there were bemused speculations as to why the process was taking so long), as well as repeated warnings by Hungarian centrist politicians and by the clergy that pressure on them was increasing and in the end would prove irresistible. These reports, and the perceptions that underlie them, suggest what Ferenc Nagy called a "master plan" and the local variations in its execution do not negate its existence.

Still, much remains unexplained by this scenario. What some dismiss as tactical restraint, or local variants, in fact had a life of their own that is hard to fit into a premeditated plan of sovietization. Tactical restraint was bad policy; far from intimidating the sturdy, self-assertive reaction, it gave it an opportunity to enlist broad popular support for what most perceived, correctly, as an antisoviet position. It was also likely to confuse Hungary's neighbors, all of them closer to soviet hearts than the obstreperous Magyars, and cause them to wonder why a nation with such a dismal past should be allowed a pluralistic state system while they were being mercilessly sovietized by fiat. And if the restraint was experimental, what was it supposed to discover? The political temper of Hungarians had been on display for centuries with few if any redeeming features; what chance was there that it would reform on its own?

The essential facts are indisputable: After the country's liberation the soviet leadership invited, accommodated, and supported a multiparty coalition government whose task it was to usher the country from a state of war (in which part of the nation was still engaged) to a rather lenient armistice. Furthermore it didn't merely tolerate the presence of reactionary elements in this government but actually insisted on it—and at the same time deliberately excluded its most trusted agents (the "exchanged ones"). It curbed the sanguine elements of the emerging Communist Party who demanded instant sovietization, went to great lengths to convince the populace that the Red Army had come to Hungary as liberators with no vengeance against the defeated nation, and denied outright support to the Communist Party, allowing it to fight its own battles. One of those battles was the quest for control of the police, a question on which Rákosi pledged not to compromise, yet when the battle threatened the survival of the coalition the soviets made it clear that they would not tolerate such an outcome, and the ratio of police officers was reapportioned in favor of the Smallholders.

The elections of November 1945 are the most telling evidence in the "special case" theory. In the postwar history of East Europe those elections were unique and beyond facile rationalizations. The soviets had potent means to influence the outcome of any election, plebiscite, or referendum. In the immediate postwar years they seemed to perceive no need for any election at all in Poland, Rumania, and Bulgaria, countries already integrated into the soviet bloc. However in Hungary they passively observed as the reaction scored a sweeping victory; their press even commented favorably on both the elections and the outcome. Two years later, with the Hungarian Communist Party (MKP) well entrenched and having the means of enforcement at its disposal, the elections, though marred by abuses, were still relatively free, but the fragmentation of the opposition allowed the communists to emerge as the largest party, with less than a quarter of the votes cast.

Rákosi's rather fanciful assumption that Hungary's sovietization would be delayed until a new, and possibly final, crisis erupted in the west might have had some polemical value but it did not explain why that outcome had to be awaited in the case of Hungary and not of the other satellites. Sovietization, whether by premeditated or opportunistic means, is an internal process; its effects nevertheless are powerfully reflected in the conduct of foreign policy. Internal and external affairs cannot for long be based on different or contrary premises. This is why the diplomacy of East European nations after World War II is of special interest. Where Hungary was concerned, this study conclusively shows that, in the first three years at any rate, the soviets exerted very little influence on Hungarian foreign policy, or indeed on internal affairs. It was all of a piece of course: had the MKP, with soviet support, established its authority from the beginning, the foreign ministry and the diplomatic service would not have remained in the hands of the Smallholders and of the Old Guard respectively for so long and their prowestern orientation would have been severely curbed. Not until the debate over the Marshall Plan was some soviet pressure exerted and even then it took the modest form of the soviet envoy explaining his government's negative position. There was nothing to prevent the cabinet from voting in favor of sending a delegation to Paris; the soviets might have expressed displeasure as they had the year before when Ferenc Nagy led a government delegation to the western capitals, but almost certainly would not have gone beyond that. In this case, as in others, it was self-induced psychological pressure rather than actual coercion that stayed the government's hand.

The argument that the communists allowed the Smallholders to direct foreign affairs in these early years because they wanted to exempt themselves from the peace process, which was likely to have bad results, does not stand up to examination. It was not the conference in Paris that made the peace and the sophisticated Hungarian public was well aware of that. All key decisions were entrusted to the Council of Foreign Ministers and even the final resolutions of the peace conference were subject to its ultimate approval. So far as the deliberations of the council were concerned it was an open secret that on all questions of great concern to Hungary the soviets took an inimical position: they opposed any change in the Transylvanian border, they endorsed the Czechoslovak plan to forcibly resettle the Hungarian minority, and they vetoed the formation of a tripartite commission to devise means for the solution of Hungary's economic crisis—in Hungarian minds the MKP had to share the onus of these adverse decisions. Thus nothing was gained by having a Smallholder put his hand to another unhappy peace treaty.

The communists, apart from advocating an unswervingly prosoviet stance, did not have a foreign policy prescription. Their internal memos stressed the need for a diplomacy in the hands of the proletariat but they did not explain how this was to be achieved. (Apart from all else, it was a foregone conclusion that once a proletarian graduated into the elite ranks of diplomats he would cease to be a proletarian and, with a loss of class-consciousness, would become politically irrelevant.) In any case, Hungary in the early postwar years could scarcely afford to be served by an inept and amateurish diplomatic establishment that was guided by ideology rather than pragmatism. Thus the concept—and practice—of proletarian diplomacy by whatever definition was for now set aside.

In most important capitals the "new" Hungary was represented by the Old Guard. Despite occasional annoyed comments on this state of affairs, no change was being contemplated—in fact most of the new appointments were made from those ranks. Aladár Szegedy-Maszák in Washington, László Bede in London, Pál Auer in Paris, and Gyula Szekfü in Moscow might have had some vague liberal leanings but they had very little in common with the new Hungary.

The first turning point came with Ferenc Nagy's removal from the premiership and the spate of resignations that followed. The foreign ministry was totally unprepared for such a mass exodus of professional diplomats: that the MKP too was caught off guard is evident from the fact that for some three years legations abroad were in a state of prolonged transition, operating (except perhaps in Moscow) with an incompetent skeleton personnel that was incapable not only of representing democratic Hungary in a dignified manner but even of carrying on the day-to-day business of the legations. The fact that later appointments came largely from Marxist ranks was due not to a deliberate policy of proletarianizing diplomacy but to the refusal after the Nagy affair of most trained diplomats of the old school to serve the new regime.

The second great turning point was occasioned by the Tito affair. Even a cursory review of ministry dispatches, diplomatic reports, and public pronouncements, as well as internal communications, shows a sharp and terrified shift to the left. This was the beginning of the end of Stalinism and, given Stalin's temperament, that was just what made it so serious. Beginning in the summer of 1948 the question of what political course Stalin had intended for Hungary was of no further importance. Having condemned Josip Broz Tito and Titoism with a vehemence and venom unusual even for him, Stalin laid down the rules of foreign policy in his sphere with a firmness that left no room for individual variations. The charge of Titoism hung like a sword of Damocles over the head of every political figure of whatever antecedents; the autonomy of Hungarian foreign policy came to an end—but it could never be shown that this was a phase in a deliberate process. Nothing in fact had seemed more unlikely in 1945, 1946, and even 1947, than the break between Tito's party and the rest of the soviet bloc that burst upon the world in the summer of 1948.

Thus the question of what fate Stalin contemplated for Hungary must be applied to the period before 1948. Within that context the most persuasive answer is that Hungary was a special case, a laboratory of democracy from above, to coin a new phrase. Of all the East European countries it had the sturdiest and most cultured middle class, which also appeared to be politically pliable. Stalin might have harked back to the urgings of the prerevolutionary intelligentsia to enlist the middle class, with its capacity for transient leadership, in the revolutionary struggle. The great number of Jews in Communist Party ranks was a promising sign as Jews had long held a commanding position in the nation's cultural life. The working class and the peasants on the other hand had been generally apathetic.

What Stalin obviously didn't reckon with was the strength and staying power of the reaction in Hungary. He must have expected, not unreasonably, that the catastrophes into which the regimes dominated by it had carried the nation would turn the masses against it. On the foreign policy scene he must have been aware that the chief features of reactionary rule—the dominance of the agrarian elite, the association between church and state, the abiding anti-Semitism—would be nearly as repulsive in the eyes of the western powers as they were to communists, hence cooperation between a government dominated by the Smallholders and western capitalist countries, even if it yielded temporary economic benefits, would be short-lived.

If this was indeed Stalin's reasoning there was solid logic behind it but logic seldom fares well in politics. The reaction was indeed successfully discredited but that did not translate into communist gains. The disillusionment of middle-class intellectuals with communist rule came as swiftly as had their initial alignment with it. And, given Cold War realities, the western powers were ready to embrace any political group regardless of its unattractive features as long as it was steadfastly anticommunist.

Hungary was sovietized, by dint not of persuasion but of terror. But, unlike the terror in Stalin's Russia that numbed all opposition, the terror in Hungary invigorated its critics: that narrow intellectual echelon carried the whole nation with it. In the Revolution of 1956 the "providential men" were swept away and new men, drawn largely from among their victims, took their place. They instituted a system of "velvet communism" that made the transition to a more open system smoother and less catastrophic. The one clear gain was the demise of the reaction. No feature of it survived; there are no ballasts to hold back progress toward a still newer Hungary.

The larger question that preoccupies many minds is whether the political faultline that runs through East Europe has corrected itself (with Yugoslavia alone subject to its tremors) and whether the region can look forward to a more stable future. For too long the fate of East Europe was determined by the flanking powers, Germany and Russia; the region enjoyed a fragile independence only when these powers were weakened by dissension or laid low by defeat. Resurgence on the wings, in the east and the west, invariably produced attempts to envelop the region. At present the wing powers are strategically helpless. Whether this signals an end to their imperialistic impulses and a cancellation of their territorial claims is a question of the foremost importance for the fate of East Europe.

Notes

LIST OF ABBREVIATIONS USED IN THE NOTES:

British Overseas	*Documents on British Foreign Policy Overseas*
Rákosi	"Dokumentek Rákositol Rákosirol"
FRUS	*Foreign Relations of the United States*
Hungary-Czechoslovakia	*Hungary and the Conference of Paris: Hungarian-Czechoslovak Relations. Papers and Documents Relating to the Preparation of Peace and the Exchange of Populations between Hungary and Czechoslovakia*
IPH	Institute Political History
Western Visit	"Iratok a Nagy Ferenc vezette magyar kormanykuldottseg 1946 evi amerikai latogatasanak tortenetehez"

Notes to Chapter 1

1. Gyula Juhász, *Magyarország külpolitikája, 1919-1945* (Hungarian Foreign Policy, 1919-1945) Budapest: Kossuth Kiadó, 1988, 452-53.

2. Nicholas Kállay, *Hungarian Premier: A Personal Account of a Nation's Struggle in the Second World War*, Westport, Conn.: Greenwood Press, 1964, 370.

3. Ibid., 374.

4. Institute for Political History (Hereafter, IPH), Budapest, Social Deomcratic Party (SDP) Files, Fond 283, Foreign Policy & Nationalities Bureau Documents, 12/122.

5. Cited in, Mihály Korom, *A Magyar fegyverszünet, 1945* (The Hungarian Armistice, 1945). Budapest: Kossuth Kiadó, 1987, 9.

6. *Foreign Relations of the United States* (Hereafter, FRUS) Washington, D. C.: Government Printing Office, 1944, Vol. III, 847.

7. Kállay, 397.

8. FRUS, 1944, III, 850-51.

9. Ibid., 847.

10. *Horthy Miklós Titkos Iratai* (Secret Papers of Miklós Horthy), Miklós Szinai and László Szücs, eds. Budapest: Kossuth Kiadó, 1962, 298.

11. Ibid., 472.

12. Kállay, 378.

13. Mihály Korom, *Magyarország ideiglenes kormánya és a fegyverszünet, 1944-45* (Hungary's Provisional Government and the Armistice, 1944-45). Budapest: Akadémiai Kiadó, 1981, 104, fn.

14. Ibid., 108, fn.

15. Korom, Magyar Fegyverszünet, 21.
16. Ibid.
17. IPH, Hungarian Communist Party (MKP) Files. Central Leadership Documents, Fond 7/1.
18. Ibid., Fond 7/4.
19. Ibid.
20. *A Wilhelmstrasse és Magyarország: Német Diplomáciai Iratok Magyarországról, 1933-1944* (The Wilhelmstrasse and Hungary: German Diplomatic Papers Relating to Hungary, 1933-1944). Budapest: Kossuth Kiadó, 1968, 910.
21. Ibid., 911.
22. Ibid., 907.

Notes to Chapter 2

1. Gerö's notes in Russian are in IPH, Hungarian Communist Party (MKP) Files, Fond 274, 7/8. They are excerpted in Korom, Magyarország Ideiglenes Kormánya, 326 ff.
2. Korom, Magyarország Ideiglenes Kormánya, 330.
3. Ibid., 329.
4. Ibid.
5. Korom, Fegyverszünet, 82-83.
6. IPH, MKP Files, Central Leadership Documents, Fond 7/1.
7. IPH, MKP Files, Fond 274, 911.
8. Quoted in Korom, Magyarország Ideiglenes Kormánya, 87.
9. Sándor Balogh, *Magyarországg külpolitikája, 1945-1950* (Hungarian Foreign Policy, 1945-1950). Budapest: Kossuth Kiadó, 1988, 6.
10. FRUS, 1944, III, 936-97.
11. Ibid., 938.
12. Ibid.
13. Ibid., 952
14. Ibid., 946-47.
15. Ibid., 943-44.
16. Balogh, 15.
17. FRUS, 1944, III, 967-68.
18. Balogh, 15.
19. FRUS, 1944, III, 956-62.
20. *Churchill and Roosevelt: The Complete Correspondence*, Warren F. Kimball, Ed. Princeton: Princeton University Press, 1984, Vol. III, 351.
21. New Hungarian Central Archives (Hereafter, NHCA), Budapest, Foreign Ministry Files, political, Box 172, 9/e.
22. Harold Nicolson, *Peacemaking, 1919.* New York: Grosset & Dunlap, 1965, 126-27.

23. Hungarian National Archives, (Hereafter, NHA), Papers of Miklós Kozma, File 27.

Notes to Chapter 3

1. Pál Auer, *Fél évszázad, Események, emberek* (Half a Century, Events and People). Cited in Balogh, 32.

2. IPH, Social Democratic Party (SDP) Files; Foreign Policy and Nationalities Bureau Documents, Fond 283, 12/122.

3. NHCA, Foreign Ministry Files, Foreign Administrative Mixed, Box 1, 1/a.

4. Ibid.

5. NHCA. Foreign Ministry Files, Hungarian-Soviet Relations, Box 5, IV/100-2.

6. Ibid.

7. Ibid.

8. Ibid.

9. Ibid.

10. NHCA, Foreign Ministry Files, Soviet Union, Secret, Box 23, IV/482.

11. Ibid.

12. Ibid.

13. Ibid.

14. Ibid.

15. IPH, Hungarian Communist Party (MKP) Files, Central Leadership Documents, Fond 276,7/13.

16. Ibid., 7/16.

17. "Dokumentek Rákositol Rákosirol," (Documents from Rákosi on Rákosi)(Hereafter, Rákosi), *Multunk*, Vol. xxxvi, Nos. 2 and 3, 246.

18. Ibid., 247.

19. IPH, MKP Files, 7/34.

20. Balogh, 29.

21. Ibid., 29-30.

22. Rákosi, 249.

23. NHCA, Foreign Ministry Files, Hungarian-Soviet Relations, Box 5 IV/10-2.

24. Ibid.

25. Quoted in, Mihály Ruff, "A demokratikus magyar külpolitika kialakulása, 1945-47" (The Evolvement of a Democratic Hungarian Foreign Policy, 1945-47), in *Elméleti és módszertani közlemények*, Budapest, 1970, 54.

26. IPH, MKP Files, Fond 274, 11/84.

27. Rákosi. 250-51.

28. Stanley M. Max, *The United States, Great Britain and the Sovietization of Hungary*. New York: Columbia University Press, 1985, 10.

29. Rákosi, 246-7.

30. IPH, SDP Files, Foreign Policy and Nationalities Bureau Documents, 12/1.

31. Ibid.

32. NHCA, Foreign Ministry Files, Hungarian-Soviet Relations, IV/ 100-2.

33. NHCA, Foreign Ministry Files, Foreign Administrative, Mixed, Box 13, 4/ca.

34. Ibid., Box 171, 9/e.

Notes to Chapter 4

1. Vladimir Clementis, "The Czech-Magyar Relationship," in *Central European Observer*, No. 1, 1943, 69.

2. *Czechoslovak*, London, January 15, 1943.

3. Cited in *Szabad Nép*, Budapest, December 15, 1945.

4. Ruff, 55.

5. FRUS, 1945, IV, 422.

6. Ibid., 427.

7. Ibid., 429.

8. Ibid., 428.

9. Cited in Balogh, 40.

10. Ibid., 106-7.

11. Ibid., 104.

12. Ibid., 31.

13. IPH, Hungarian Communist Party (MKP) Files, Political Committee Minutes, MKP, Fond 274, June 7, 1945.

14. Ibid.

15. FRUS, 1945, IV, 431.

16. IPH, MKP Files, Political Committee Minutes, Fond 274, August 2, 1945.

17. Ibid., June 5, 1945.

18. Rákosi, 283.

19. Ibid., 275-76.

Notes to Chapter 5

1. FRUS, 1945, IV, 814.

2. As reported in *Szabad Nép*, Budapest, April 18, 1945.

3. IPH, Social Democratic Party (SDP) Files, Foreign Policy and Nationalities Bureau, Documents, Fond 284, Group XIII.

4. NHCA, Foreign Ministry Files, Hungarian-Soviet Relations, Secret, Box 13, IV/264.

5. NHCA, USA Administrative, Part I, Box 5, b/b.

6. IPH, Hungarian Communist Party (MKP) Files, Central Leadership Documents, Fond 274, 7/247.

7. NHCA, Foreign Ministry Files, Hungarian-Soviet Relations, Secret, Box 13, IV/264.

8. NHCA, Foreign Ministry Files, Foreign, Box 172, 9/c, 1596.

Notes to Chapter 6

1. FRUS, 1945, IV, 818.
2. IPH, Hungarian Communist Party (MKP) Files, Fond 274, 10/80.
3. Péter Sipos and István Vida, "Iratok a magyar-amerikai diplomáciai kapcsolatok 1945 evi ujrafelvételéhez" (Documents on the Resumption of Diplomatic Relations between Hungary and the United States in the Year 1945), *Századok*, December, 1987, No. 3, 421.
4. FRUS, 1945, IV, 808.
5. Sipos and Vida, 422.
6. FRUS, 1945, IV, 819.
7. Ibid., 820.
8. Ibid., 820-21.
9. Ibid., 547-48.
10. Ibid., 548.
11. Ibid., 815.
12. Ibid., fn 40.
13. Sipos and Vida, 428.
14. FRUS, 1945, IV, 815.
15. Ibid., 550.
16. Ibid., fn.
17. Ibid., 554-55.
18. *Documents on British Foreign Policy Overseas* (Hereafter, British Overseas). London: Her Majesty's Stationery Office, 1949. Series I, Vol. I, 57, fn.
19. Ibid., 3.

Notes to Chapter 7

1. For a discussion of East Central Europe's economic plight after World War I see Magda Adám, *Magyarország és a Kisantant a Harmincas Években* (Hungary and the Little Entente in the 1930s). Budapest: Akadémiai Kiadó, 1968, 41-54.
2. In FRUS, 1944, IV, 958.
3. NHCA, Foreign Ministry Files, Rumania, Secret, Box 16, 4/bc.
4. Balogh, 67.
5. Ibid.
6. IPH, Social Democratic Party (SDP) Files, Foreign Policy and Nationalities Bureau Documents, Fond 283, 1217, 42.
7. NHCA, Foreign Ministry Files, Hungarian-Soviet Relations, Administrative, Box 5, IV/100-2.
8. FRUS, 1945, IV, 516-17.
9. NHCA, Foreign Ministry Files, Hungarian-Soviet Relations, IV/100-102.
10. Balogh, 67-68.

11. János Nemes, *Rákosi Mátyás Születésnapja* (Mátyas Rákosi's Birthday). Budapest: Láng Kiadó, 1988, 140.

12. Balogh, 71.

13. FRUS, 1945, II, 124.

14. Rákosi, 275-77.

15. IPH, Hungarian Communist Party (MKP) Files, Fond 174, 7/237.

16. Ibid.

17. Ibid., 7/216.

18. William Taubman, *Stalin's American Policy: From Entente to Cold War*. New York: W. W. Norton, 1988, 78.

Notes to Chapter 8

1. FRUS, 1945, II, Potsdam Conference, 643-44.

2. Ibid., 698.

3. NHCA, Foreign Ministry Files, Hungarian-Soviet Relations, Administrative, Box 14, IV/170.

4. NHCA, Foreign Ministry Files, Soviet Union, Secret, Box 23, IV/482.

5. NHCA, Foreign Ministry Files, Hungarian-Soviet Relations, Box 11, IV/18/1.

6. Ibid.

7. Ibid.

8. British Overseas, Series I, Vol. II, 222.

9. FRUS, 1945, II, 182-83.

10. Ibid.

11. British Overseas, Series I, Vol. I, 112.

12. Ibid., I, II, 259-60.

13. Ibid.

14. NHCA, Foreign Ministry Files, USA Administrative, Box 18.

15. Balogh, 54.

16. Ibid.

17. FRUS, 1945, IV, 828-29.

18. Balogh, 54.

19. Ibid., 55.

20. NHCA, Foreign Ministry Files, Hungarian-British Relations, Administrative, 4/b.

21. FRUS, 1945, IV, 847.

22. Ibid., 848-50.

23. Ibid., 851.

24. IPH, Social Democratic Party (SDP) Files, Foreign Policy and Nationalities Bureau Documents, Fond 283, 12-2.

25. Ibid.

26. IPH, Hungarian Communist Party (MKP) Files, Political Committee Minutes, Fond 274, October 8, 1945.

27. IPH, Social Democratic Party (SDP) Files, Foreign Policy and Nationalities Bureau Documents, Fond 283, 12-1.

28. Balogh, 57.

29. Ibid., 57-58.

30. FRUS, 1945, IV, 866-67.

31. Ibid., 874, fn. 81.

32. NHCA, Foreign Ministry Files, USA Administrative, Box 11, 5/bc.

33. FRUS, 1945, IV, 848.

34. Ibid., 876.

35. IPH, SDP Files, Foreign Policy and Nationalities Bureau Documents, Fond 283, 12-1.

36. FRUS, 1945, IV, 886.

37. Ibid., 868-69.

38. Ibid., 869 fn.

39. Ibid., 887.

40. Ibid., 888.

41. Ibid., 882.

42. Ibid., 898.

Notes to Chapter 9

1. FRUS, 1945, IV, 891-92.

2. IPH, Hungarian Communist Party (MKP) Files, Documents on British and U.S. Missions, Fond 274, 10/80.

3. Quoted in *Kis Ujság*, Budapest, November 6, 1945.

4. NHCA, Foreign Ministry Files, Foreign, Box 1, 1/a.

5. FRUS, 1945, IV, 901.

6. Ibid., 922-23.

7. Quoted in *Szabad Nép*, Budapest, December 21, 1945.

8. Max, 66.

9. *Hungary and the Conference of Paris: Hungarian-Czechoslovak Relations. Papers and Documents Relating to the Preparation of Peace and the Exchange of Populations Between Hungary and Czechoslovakia.* (Hereafter, Hungary-Czechoslovakia). Budapest: Hungarian Ministry of Foreign Affairs, 1947, 18.

10. Ibid., 20.

11. Ibid., 26.

12. Balogh, 106.

13. FRUS, 1945, IV, 935-36.

14. Ibid., 946.

15. Balogh, 112.

Notes to Chapter 10

1. Rákosi, 275.

2. Ibid., 278, fn.

3. Geoffrey and Nigel Swain, *Eastern Europe since 1945*, New York: St. Martin's Press, 1993, 51.

4. NHCA, Foreign Ministry Files, Foreign, Box 171, 9/e.

5. NHCA, Foreign Ministry Files, Soviet Union, Secret, Box 22, IV/482.

6. NHCA, Foreign Ministry Files, Soviet Union and Yugoslavia, Secret, Box 6, IV/103.

7. Ruff, 58.

8. *Igazság*, Budapest, January 8, 1946.

9. Balogh, 143.

10. Ibid., 114.

11. NHCA, Parliamentary Documents, Foreign Policy Committee Minutes, Box 10, 4/cc.

12. Balogh, 151.

13. NHCA, Foreign Ministry Files, Foreign, Administrative, Mixed, Box 10.

14. Balogh, 143.

15. FRUS, 1946, VI, 580.

16. Ibid., 581.

17. Ibid., 272-73.

18. Balogh, 147.

19. *CAS*, Prague, July 3, 1945.

20. Zsuzsa L. Nagy, "Az 1920-as magyar békeszerződés" (The 1920 Hungarian Peace Treaty), *Propagandista*, 1980, No. 2, 110-13.

21. *CAS*, Prague, July 3, 1945.

22. Ferenc Krasovec, "A magyar köztársaság kikiáltása" (The Proclamation of the Hungarian Republic), in *Tanulmányok a magyar népi demokratikus forradalom történetéböl* (Studies in the History of the Hungarian People's Democratic Revolution). Budapest: Hungarian Scientific Academy, 1966, 123.

23. Ibid.

24. Quoted in Attila Király, "A Nagy Ferenc kormány külpolitikai törekvései és a parlament" (Foreign Policy Endeavors of the Ferenc Nagy Government and Parliament), in ibid., 146.

25. Ibid.

26. FRUS, 1946, VI, 258-9.

27. Ibid., 256.

28. Ibid., 251.

29. Ibid., 252.

30. Ibid., 251.

Notes to Chapter 11

1. Ferenc Nagy, *The Struggle Behind the Iron Curtain.* New York: Macmillan, 1948, 208-9.

2. NHCA, Parliamentary Documents, Foreign Policy Committee Minutes, Box 10, 4cc.

3. Nagy, *Struggle*, 209.

4. Ibid.

5. Ibid., 208.

6. FRUS, 1946, VI, 282.

7. NHCA, Foreign Ministry Files, London Embassy Reports, Box 1, 1/a.

8. NHCA, Parliamentary Documents, Foreign Policy Committee Minutes, Box 11, III/8.

9. Balogh, 179.

10. NHCA, Foreign Ministry Files, Rumania, Administrative, Box 5, 4/bc.

11. Balogh, 180.

12. Mihály Fülöp, "A külügyminiszterek tanácsa és a magyar békeszerzödés" (The Council of Foreign Ministers and the Hungarian Peace Treaty), *Külpolitika*, 1985, No. 4, 139.

13. Balogh, 180.

14. Ibid., 181.

15. NHCA, Foreign Ministry Files, USA Administrative, Box 22, 1369 political.

16. FRUS, 1946, VI, 282-83.

17. Ibid.

18. Ibid.

19. Balogh, 174.

20. NHCA, Foreign Ministry Files, Rumania, Administrative, Box 5, 4/bc.

21. NHCA, Foreign Ministry Files, Rumania, Secret, Box 16.

22. Mária Koncz interview with Wladimir Magyari-Block, *Magyar Nemzet*, Budapest, March 18, 1989.

Notes to Chapter 12

1. Nagy, *Struggle*, 222-3.

2. Max, 81.

3. "Iratok a Nagy Ferenc vezette magyar kormányküldöttség 1946 évi amerikai látogatásának történetéhez" (Documents on the 1946 Visit of the Hungarian Government Delegation led by Ferenc Nagy to America) (hereafter Western Visit), in *Levéltári közlemények*, Budapest, 1978, 245.

4. FRUS, 1946, VI, 287-89.

5. Western Visit, 249.

6. Ibid.

7. Ibid.

8. Nagy, *Struggle*, 225.

9. Ibid., 223.

10. Western Visit, 250.

11. Ibid.

12. Ibid., 251.

13. Ibid.

14. Ibid., 260.

15. Ibid., 261.

16. Ibid.

17. Ibid., 252.

18. Ibid.

19. Ibid., 253.

20. FRUS, 1946, VI, 302-4.

21. Ibid., 308-9.

22. Ibid., 311.

23. Ibid., 315.

24. Ibid., 314-16.

25. Nagy, *Struggle*, 232.

26. NHCA, Parliamentary Documents, Parliamentary Foreign Policy Committee Minutes, Box 10, 4/cc.

27. Nagy, *Struggle*, 236.

28. Ibid., 237.

29. Western Visit, 256; IPH, Hungarian Communist Party (MKP) Files, Fond 274, 7/89.

30. NHCA, Foreign Ministry Files, Hungarian-Soviet Relations, Box 5, IV/101.

31. Ibid.

32. Nemes, 112.

33. Western Visit, 256.

Notes to Chapter 13

1. Balogh, 121.

2. NHCA, Foreign Ministry Files, USA Administrative, Box 11, 4/ba.

3. NHCA, Foreign Ministry Files, Soviet Union-Yugoslavia, Secret, Box 6, IV/103.

4. NHCA, Parliamentary Documents, Parliamentary Foreign Policy Committee Minutes, Box 10, July 9, 1946.

5. Balogh, 189.

6. Ibid., 190.

7. Ibid., 123.

8. Ibid., 212.

9. IPH, Social Democratic Party (SDP) Files, Fond 283, 12/101.

10. Taubman, 132-33.

11. NHCA, Foreign Ministry Files, Hungarian-Soviet Relations, Secret, Box 5, 11/8.

Notes to Chapter 14

1. NHCA, Foreign Ministry Files, USA Administrative, Box 171, 9/e.

2. Balogh, 214.

3. NHCA, Foreign Ministry Files, USA Administrative, Box 22, 2018.

4. NHCA, Parliamentary Documents, Parliamentary Foreign Policy Committee Minutes, August 3 and 5, 1946.

5. FRUS, 1946, III, 86.

6. Ibid., 122-23.

7. Ibid., 213.

8. Ibid., 210ff.

9. Ibid., 215ff.

10. Ibid., 257-59.

11. Ibid., 276-77.

12. Ibid. 330-31.

13. Ibid., 339.

14. Ibid., 370-72.

15. Ibid., 332-33.

16. NHCA, Parliamentary Documents, Parliamentary Foreign Policy Committee Minutes, September 9, 1946.

17. Ibid.

18. *Szabad Nép*, Budapest, September 7, 1946.

19. As reported in *Pravda*, Moscow, September 22, 1946.

20. Balogh, 239.

21. IPH, Social Democratic Party (SDP) Files, Foreign Policy and Nationalities Bureau Documents, Fond 283, 12/134.

22. FRUS, 1946, IV, 531.

23. Ibid., 539.

24. As reported in *Kis Ujság*, Budapest, October 4, 1946.

25. As reported in *Pravda*, Moscow, October 4, 1946.

26. FRUS, 1946, VI, 258-59.

27. For the text of the peace treaty, see "Hungarian Peace Treaty, Paris, February 10, 1946," in *Major Peace Treaties of Modern History: 1648-1967*, Fred Israel, ed., New York: Chelsea House Publishers, 1967, Vol. IV, 2553-83.

28. NHCA, Parliamentary Documents, Parliamentary Foreign Policy Committee Minutes, November 8, 1946.

Notes to Chapter 15

1. NHCA, Foreign Ministry Files, Hungarian-British Relations, Administrative, Box 9, 4/b.

2. NHCA, Parliamentary Documents, Parliamentary Foreign Policy Committee Minutes, November 8, 1947.

3. NHCA, Foreign Ministry Files, Foreign Mixed, Box 171, 9/e.

4. FRUS, 1946, VI, 345fn.

5. Ibid., 346-47.

6. Ibid., 345.

7. IPH, Hungarian Communist Party (MKP) Files, Political Committee Minutes, Fond 274, December 4, 1946.

8. NHCA, Foreign Ministry Files, Hungarian-Soviet Relations, Fond 283, 12/119.

9. FRUS, 1947, IV, 261.

10. Ibid., 263.

11. Ibid., 264.

12. IPH, MKP Files, Political Committee Minutes, Fond 274, January 30, 1947.

13. Ibid., January 20, 1947.

14. NHCA, Parliamentary Documents, Parliamentary Foreign Policy Committee Minutes, January 30, 1947.

15. Balogh, 248.

16. Ibid., 230-31.

17. FRUS, 1947, IV, 266.

18. Balogh, 129.

19. FRUS, 1947, IV, 268-69.

20. Ibid., 260.

21. Ibid.

22. Ibid., 272-73.

23. Ibid., 274-75.

24. Ibid., 277-78.

25. Ibid., 278-79.

26. Ibid., 285-86.

27. IPH, MKP Files, Politicsl Committee Minutes, Fond 274, April 3, 1947.

28. IPH, MKP Files, Central Leadership Documents, Fond 274, 7/158.

29. NHCA, Foreign Ministry Files, USA Administrative, Box 22, 2045.

30. Nagy, *Struggle*, 373.

31. Max, 96. Max is in error when he writes, on page 97, that Kovács would never be seen again and is presumed to have died in soviet captivity. He was sentenced to 20 years at forced labor but was turned over to Hungarian authorities, who released him in the fall of 1955. He was briefly minister of agriculture during the October 1956 revolution and died in 1959.

32. Nemes, 64-65.

33. NHCA, Foreign Ministry Files, Hungarian-Soviet Relations, Secret, Box 5.

34. NHCA, Foreign Ministry Files, USA Administrative, Box 22, 1369.

35. Balogh, 258.

36. IPH, Social Democratic Party (SDP) Files, Fond 243, 3/80.

37. NHCA, Foreign Ministry Files, USA Administrative, Box 16, 5/b.

38. FRUS, 1947, IV, 181-82.

39. NHCA, Foreign Ministry Files, USA Administrative, Box 16, 5/b.

40. FRUS, 1947, IV, 298.

41. Ibid., 99fn.

42. NHCA, Cabinet Minutes, May 28, 1947.

43. IPH, MKP Files, Political Committee Minutes, Fond 274, May 29, 1947.

44. FRUS, 1947, IV, 300.

45. Ibid.

46. NHCA, Foreign Ministry Files, Hungarian-Soviet Relations, Secret, Box 13, IV/264.

Notes to Chapter 16

1. NHCA, Foreign Ministry Files, Foreign Administrative Mixed, Box 11, 4/ca; FRUS, 1947, IV, 301-3.

2. FRUS, 1947, IV, 304fn.

3. Ibid., 312.

4. Balogh, 262.

5. FRUS, 1947, IV, 314.

6. Ibid., 315.

7. Ibid., 320.

8. As reported in *Kis Ujság* (organ of the Smallholders' Party), Budapest, June 19, 1947.

9. Journal of the National Assembly, Budapest, June 25, 1947, Vol. VIII, 30.

10. For a detailed account of the debate, see Balogh, 263-68.

11. NHCA, Foreign Ministry Files, USA Administrative Box 11, 1/ba.

12. FRUS, 1947, IV, 238-39.

13. NHCA, Foreign Ministry Files, USA Administrative, Box 11, 1/ba.

14. IPH, Hungarian Communist Party (MKP) Files, Central Leadership Documents, Fond 274, 7/228.

15. Ibid., 2741-47/122.

16. FRUS, 1947, IV, 238-39.

17. Ibid., 351.

18. NHCA, Foreign Ministry Files, Foreign Mixed, Box 1, 1/a.

19. FRUS, 1947, IV, 219.

20. IPH, Social Democratic Party (SDP) Files, Hungarian-Soviet Relations, 283.

21. NHCA, Council of Ministers Minutes, July 1, 1947.

22. Ibid.

23. FRUS, 1947, IV, 340.

24. NHCA, Foreign Ministry Files, English, Administrative, Box 9.

25. IPH, Hungarian Communist Party (MKP) Files, Central Leadership Documents, Fond 274, 7/237.

26. NHCA, Foreign Ministry Files, English, Administrative, Box 9.

27. Ibid.

Notes to Chapter 17

1. NHCA, Foreign Ministry Files, English, Administrative, Box 9.

2. Ibid., Box 8, 4/b.

3. IPH, Hungarian Communist Party (MKP) Files, Central Leadership Documents, Fond 274, 7/237.

4. Ibid.

5. FRUS, 1947, IV, 369-70.

6. Ibid.

7. Balogh, 270.

8. Journal of the National Assembly, Budapest, September 16, 1947, Vol. I, 109.

9. FRUS, 1947, IV, 383.

10. Ibid., 392.

11. Ibid., 391.

12. Milovan Djilas, *Conversations with Stalin*, New York: Harcourt, Brace, 1962, 155. Zhdanov remarked to Djilas how punctual and fastidious the Finns were in their deliveries. "Everything on time, everything packed, and of excellent quality."

13. IPH, MKP Files. Documents on British and U. S. Missions, Fond 247, 10/113.

14. *Szabad Nép*, Budapest, September 12, 1947.

15. IPH, Hungarian Workers' Party (MDP) Files, Secret Papers of Mátyás Rákosi, Fond 276, 65/105.

16. IPH, MDP Files, Rákosi-AVH (State Defense Authority) Correspondence, Fond 276.

17. IPH, MDP Files, Secret Papers of Mátyás Rákosi, Fond 274.

18. NHCA, Foreign Press Department Files, Yugoslavia, Box 6, 4/bc.

19. FRUS, 1948, IV, 295.

20. Balogh, 284.

21. IPH, Social Democratic Party (SDP) Files, Hungarian-Soviet Relations, Fond 293.

22. Balogh, 286-88.

23. Balogh, 289-90.

24. NHCA, Foreign Ministry Files, Foreign Mixed, Box 170, 9/e.

25. IPH, MKP Files, Central Leadership Documents, 7/122.

Notes to Chapter 18

1. Balogh, 131.
2. NHCA, Foreign Ministry Files, Moscow Embassy Reports, Mixed, Box 5.
3. FRUS, 1948, IV, 733.
4. IPH, Hungarian Communist Party (MKP) Files, Central Leadership Documents, Fond 274, Meeting of October 9, 1947.
5. Swain and Swain, 59.
6. FRUS, 1948, IV, 749-50.
7. IPH, Hungarian Workers' Party (MDP) Files, Secret Papers of Mátyás Rákosi, Fond 276, 65/120.
8. Ibid., Fond 274, 3/121.
9. FRUS, 1948, IV, 528-29.
10. Cited in Balogh, 295.
11. Ibid.
12. FRUS, 1948, IV, 311.
13. Ibid., 311-12.
14. Ibid., 317.
15. NHCA, Foreign Ministry Files, Hungarian-Soviet Relations, Secret, Box 14, IV/250.
16. Ibid.
17. Cited in FRUS, 1948, IV, 327.
18. IPH, MKP Files, Documents on USA and British Missions, Fond 274, 10/115.
19. FRUS, 1948, IV, 327.
20. Ibid.
21. Ibid., 336.

Notes to Chapter 19

1. IPH, Hungarian Workers' Party (MDP) Files, Secret Papers of Mátyás Rákosi, Fond 276, 65/114.
2. *Szabad Nép*, June 23, 1948.
3. Ibid., June 29, 1948.
4. Ibid., July 1, 1948.
5. IPH, MDP Files, Secret Papers of Mátyás Rákosi, Fond 276, 65/114.
6. Ibid., Fond 274, 7/114.
7. IPH, MDP Files, Central Leadership Documents, Fond 276.
8. IPH, MDP Files, Secret Papers of Mátyás Rákosi, 276, 65/103.
9. Ibid.
10. Max, 129-30
11. FRUS, 1948, IV, 366.
12. FRUS, Conference on Danube Navigation, 593-672.

13. FRUS, 1948, IV, 1079.

14. Ibid., 365ff.

Notes to Chapter 20

1. IPH, Hungarian Workers' Party (MDP) Files, Political Committee Minutes, Fond 276, 53/5.

2. Ibid.

3. Ibid.

4. IPH, MDP Files, Fond 276, 65/102.

5. Ibid., 65/202.

6. Ibid.

7. Ibid., 65/203.

8. FRUS, 1948, IV, 374.

9. Ibid., 385-87.

Notes to Chapter 21

1. IPH, Hungarian Workers' Party (MDP) Files, Secret Papers of Mátyás Rákosi, Fond 276, 65/112.

2. Ibid., 65/202.

3. NHCA, Foreign Ministry Files, Foreign Ministry Press Office Documents, Yugoslavia, Box 6, 4/bc.

4. NHCA, MDP Files, Central Leadership Documents, Fond 276, 65/207.

5. IPH, MDP Files, Secret Papers of Mátyás Rákosi, 65/103.

6. Ibid., 65/102.

7. IPH, MDP Files, Central Leadership Documents, 65/198.

8. NHCA, Foreign Ministry Files, English Administative, Box 5, 4/c.

9. FRUS, 1948, IV, 383-84.

10. Ibid., 384-85.

11. Ibid., 385 fn.

12. Ibid., 389.

13. FRUS, 1948, IV, 389-90.

14. IPH, MDP Files, Central Leadership Documents, 65/198.

15. Ibid.

16. Ibid.

17. FRUS, 1948, IV, 394.

18. Ibid.

19. Ibid., 395.

20. Ibid., 1949, Vol. 1.

21. As reported in *Szabad Nép*, February 16, 1949.

22. Balogh, 316.

23. FRUS, 1949, V, 452.

24. Ibid., 458.

25. Ibid.

26. Ibid., 459.

27. NHCA, Foreign Ministry Files, English Administrative, Box 9.

28. *Szabad Nép*, February 9, 1949.

29. FRUS, 1949, V, 463.

30. Ibid., 463-64.

31. Ibid., 464 fn.

32. Ibid.

33. Ibid., 465.

34. Ibid., 467-69.

35. Ibid., 466.

Notes to Chapter 22

1. IPH, Hungarian Communist Party (MKP) Files, Secret Papers of Mátyás Rákosi, Fond 276, 65/112.

2. Ibid., 65/202.

3. Balogh, 314.

4. NHCA, Foreign Minisntry Files, Hungarian-British Administrative, Box 5, 4/b.

5. Ibid.

6. Balogh, 314.

7. Ibid., 315.

8. FRUS, 1949, IV, 470.

9. NHCA, Foreign Ministry Files, Hungarian-British Administrative, Box 5, 4/b.

10. Ibid.

11. Journal of National Assembly, June 8, 1949, Vol. I, 44.

12. Ibid., 68.

13. *Szabad Nép*, June 8, 1949.

14. IPH, Hungarian Workers' Party (MDP) Files, Secret Papers of Mátyás Rákosi, 65/112.

15. IPH, MDP Files, Central Leadership Documents, Fond 276, 65/198.

16. Ibid.

17. Bits and pieces of material on the Harrison case are found in various collections. The full account is in IPH, MDP Files, Central Leadership Documents, Fond 276, 6/215.

18. NHCA, Foreign Ministry Files, English Secret, Box 5, 4/a.

19. FRUS, 1949, IV, 471-72.

20. Balogh, 329.

21. Cited in Balogh, 330.

22. NHCA, Foreign Ministry Files, Hungarian-British Administrative, Box 5, 4/b.

23. Ibid.

24. Ibid.

Notes to Chapter 23

1. NHCA, Foreign Ministry Files, Yugoslavia Administrative, Box 6, 4/bc.

2. NHCA, Foreign Ministry Press Section Documents, Yugoslavia, Box 6, 4/bc.

3. Ibid.

4. Ibid.

5. Swain and Swain, 69.

6. NHCA, Foreign Ministry Files, Foreign Mixed, Box 20, 5/c.

7. FRUS, 1949, V, 484.

8. Ibid., 483.

9. Ibid.

10. Ibid., 491.

Notes to Chapter 24

1. IPH, Hungarian Workers' Party (MDP) Files, Central Leadership Documents, Fond 276, 65/207.

2. Ibid.

3. Ibid.

4. NHCA, Foreign Ministry Files, London Embassy Reports, Box 1, 1/a.

5. Ibid.

6. IPH, MKP Files, Secret Papers of Mátyás Rákosi, Fond 276, 65/184, Rákosi-AVH.

7. FRUS, 1950, V, 982.

8. NHCA, Foreign Ministry Files, Hungarian-British Relations, Administrative, Box 8, 4/ag.

9. NHCA, Foreign Ministry Files, London Embassy Reports, Box 1, 1/a.

10. NHCA, Foreign Ministry Files, London Embassy Reports, Box 3, 3/b.

11. FRUS, 1950, IV, 994.

12. IPH, MDP Files, Secret Papers of Mátyás Rákosi, 65/198.

13. FRUS, 1950, IV, 994.

14. Ibid., 1004.

15. Ibid., 1016-17.

16. NHCA, Foreign Ministry Files, Hungarian-British Relations, Administrative, 4/ag.

17. NHCA, Foreign Ministry Files, English, Secret, Box 5, 4/a.

18. NHCA, Foreign Ministry Files, Hungarian-British Relations, Administrative, Box 8, 4/b.

Notes to Chapter 25

1. NHCA, Foreign Ministry Files, French Administrative, 4/b.

2. Ibid.

3. Ibid. 4/db.

4. Ibid., 4/b.

5. Ibid.

6. IPH, Social Democratic Party (SDP) Files, Fond 283, 12/45.

7. Ibid., 247, 10/2.

8. NHCA, Foreign Ministry Files, Hungarian-Soviet Relations, Box 5, IV/100-102.

9. Ibid.

10. NHCA, Foreign Ministry Files, Yugoslavia, Secret, Box 6, IV/103.

11. NHCA, Foreign Ministry Files, Soviet Union & Yugoslavia, Secret, Box 6, IV/103.

12. IPH, Hungarian Communist Party (MKP) Files, Documents on U.S. and British Missions, Fond 274, 10/114.

13. Ibid., 10/115.

14. NHCA, Foreign Ministry Files, Yugoslavia Administrative, Box 8.

15. Ibid.

16. NHCA, Foreign Ministry Files, Rumania Secret, Box 5, 4/bc.

17. NHCA, Rumania, Secret, Box 5, 4/bc.

18. NHCA, Rumania, Secret, Box 1.

19. Ibid.

20. Ibid.

21. Ibid.

22. Ibid.

23. Ibid.

24. Ibid.

25. Ibid.

26. Ibid.

27. NHCA, Foreign Ministry Files, USA Administrative, Box 16.

28. IPH, MDP Files, Central Leadership Documents, Fond 276, 65/198.

29. NHCA, Foreign Ministry Files, USA Administrative, Box 11, 5/bc.

30. NHCA, Foreign Ministry Files, French Administrative, 4/c.

31. IPH, MDP Files, Central Leadership Documents, 65/198.

32. NHCA, Foreign Ministry Files, USA Administrative, Box 20, 5/c.

33. IPH, MDP Files, Fond 247, 10/2.

34. Nemes, 82-83.

Notes to Chapter 26

1. IPH, Hungarian Workers' Party (MDP) Files, Secret Papers of Mátyás Rákosi, Fond 276, 65/120.

2. Ibid.

3. Ibid.

4. NHCA, Foreign Ministry Files, London Embassy Reports, Box 11, 1/a.

5. Ibid.

6. FRUS, 1949, V, 477.

7. FRUS, 1949, IV, 1022-23.

8. NHCA, Foreign Ministry Files, Hungarian-British Relations, Administrative, 4/bc.

9. FRUS, 1950, IV, 987-88.

10. NHCA, Foreign Ministry Files, Foreign Mixed, Box 20, 5/c.

11. Ibid.

12. Ibid.

13. NHCA, Foreign Ministry Files, Hungarian-Soviet Relations, 144/Conf. 1951.

Selected Bibliography

ARCHIVAL COLLECTIONS

Institute for Political History (IPH), Budapest, Hungary

Hungarian Communist Party (MKP), Hungarian Workers' Party (MDP) Files

General: Fonds 174, 274.

Central Leadership Documents: Fonds 7/1, 7/2, 7/4, 7/13, 7/16, 7/34, 274, 276

Political Committee Meetings Minutes (under Central Leadership Documents): Fond 276

Documents on USA and British Missions: Fond 247

Secret Papers of Mátyás Rákosi: Fonds 274, 276

Social Democratic Party (SDP) Files

Foreign Policy and Nationalities Bureau Documents: Fonds 243, 283, 284

Hungarian-Soviet Relations: Fond 293

New Hungarian Central Archives (NHCA) Budapest, Hungary

Cabinet Minutes

Foreign Ministry Files

Foreign

Foreign Mixed

Hungarian-British Relations, Administrative and Secret

English, Administrative and Secret

Hungarian-Soviet Relations, Administrative and Secret

Soviet Union, Administrative and Secret

USA Administrative and Secret

Rumania, Administrative and Secret

Yugoslavia, Administrative and Secret

Soviet Union-Yugoslavia, Administrative and Secret

Foreign Ministry Press Section Files

Foreign Press Department Files

PARLIAMENTARY DOCUMENTS

Journal of National Assembly

Parliamentary Foreign Policy Committee Minutes

Hungarian National Archives, Papers of Miklós Kozma, File 27

PUBLISHED PRIMARY COLLECTIONS

Documents on British Foreign Policy Overseas. London: Her Majesty's Stationery Office, Series I, Vols. I and II, 1949

Foreign Relations of the United States (FRUS). Washington, D. C.: Government Printing Office, 1944, Vol. III, 1945, Vol. IV, 1946, Vol. VI, 1947, Vol. IV, 1948, Vol. IV, 1949, Vol. V, 1950, Vol. IV

Horthy Miklós Titkos Iratai. Miklós Szinai and László Szücs, Eds. Budapest: Kossuth Kiadó, 1962.

A Wilhelmstrasse és Magyaroszág: Német Diplomáciai Iratok Magyarországról 1923-1944. Budapest: Kossuth Kiadó, 1968.

Churchill and Roosevelt: The Complete Correspondence. Warren F. Kimball, Ed. Princeton, NJ: Princeton University Press, Vol. III, 1984.

"Dokumentek Rákositól." *Multunk*, Vol. XXXVII, Nos. 2 and 3.

Hungary and the Conference of Paris: Hungarian-Czechoslovak Relations. Papers and Documents Relating to the Preparation of Peace and the Exchange of Populations between Hungary and Czechoslovakia. Budapest: Hungarian Ministry of Foreign Affairs, 1947.

"Iratok a magyar-amerikai diplomáciai kapcsolatok 1945 évi ujrafelvételéhez." Péter Sipos and István Vida, Eds. *Századok*, No. 3, 1987.

"Iratok a Nagy Ferenc vezette magyar kormányküldöttség 1946 évi látogatásának töténetéhez." *Levéltári Közlemények*. Budapest, 1978.

A Magyar Néfront Története; Dokumentumok, 1935-1976. Budapest: Kossuth Kiadó, 1972.

BOOKS

Aczél, Tamás and Mérai, Tibor. *Tisztitó Vihar*. Szeged, Hungary: Jate Kiadó, 1989.

Ádám, Magda. *Magyaroszág és a Kisantant a Harmincas Években.* Budapest: Akadémiai Kiadó, 1968.

Balogh, Sándor. *Magyaroszág Külpolitikája, 1945-1950.* Budapest: Kossuth Kiadó, 1989.

Benda, Kálmán, Ed. *Magyaroszág Történeti Kronologiája.* Budapest: Akadémiai Kiadó, 1983.

Bölönyi, József. *Magyaroszág Kormányai, 1848-1992.* Budapest: Akadémiai Kiadó, 1992.

Djilas, Milovan. *Conversations with Stalin.* New York: Harcourt-Brace, 1962.

Fejtö, Ferenc. *A Népi Demokráciák Története.* Budapest: Magvetö Kiadó, 1991.

Juhász, Gyula. *Magyarország Külpolitikája, 1919-1945.* Budapest: Kossuth Kiadó, 1988.

Kállay, Nicholas. *Hungarian Premier: A Personal Account of a Nation's Struggle in the Second World War.* Westport, Conn.: Greenwood Press, 1964.

Korom, Mihály. *A Magyar Fegyverszünet, 1945.* Budapest: Kossuth Kiadó, 1989.

———. *Magyaroszág Ideiglenes Kormánya és a Fegyverszünet, 1944-1945.* Budapest: Akadémiai Kiadó, 1988.

Kossuth Kiadó. *A Magyar Népidemokrácia Története, 1944-1962.* Budapest: Kossuth Kiadó, 1978.

Kun, József. *Hungarian Foreign Policy: The Experience of a New Democracy.* Westport, Conn.: Praeger Publishers, 1993.

Legter, Lyman H. *Eastern Europe: Transformation and Revolution, 1945-1991.* Lexington, Mass.: D. C. Heath, 1992.

Max, Stanley M. *The United States, Great Britain and the Sovietization of Hungary.* New York: Columbia University Press, 1985.

Nagy, Ferenc. *The Struggle Behind the Iron Curtain.* New York: Macmillan, 1948.

Nemes, János. *Rákosi Mátyás Születésnapja.* Budapest: Láng Kiadó, 1988.

Nicolson, Harold. *Peacemaking, 1919.* New York: Grosset & Dunlap, 1965.

Simons, Thomas W. *Eastern Europe in the Postwar World*. New York: St. Martin's Press, 1991.

Swain, Geoffrey and Nigel. *Eastern Europe since 1945*. New York: St. Martin's Press, 1993.

Taubman, William. *Stalin's American Policy: From Entente to Cold War*. New York: W. W. Norton, 1988.

Zinner, Tibor, Ed. *Rajk László és Társai a Népbiróság Elött, 40 Év Távlatából*. Budapest: Magyar Eszperanto Szövetség, 1989.

ARTICLES AND SPECIAL STUDIES

Clementis, Vladimir. "The Czech-Magyar Relationship." *Central European Observer*, No. 1, 1984.

Fülöp, Mihály. "A külügyminiszterek tanácsa és a magyar békeszerzödés." *Külpolitika*, No. 4, 1984.

Király, Attila. "A Nagy Ferenc kormány külpolitikai törekvései és a parlament." *Tanulmányok a Magyar Népidemokratikus Forradalom történetéböl*. Budapest: Hungarian Scientific Academy, 1966.

Krasovec, Ferenc. "A magyar köztársaság kikiáltása." *Tanulmányok a magyar Népi Demokratikus Forradalom Történetéböl*. Budapest: Hungarian Scientific Academy, 1966.

Nagy, Zsuzsa L. "Az 1920-as magyar békeszerzödés." *Propagandista*, No. 2, 1980.

Ruff, Mihály. "A demokratikus magyar külpolitika kialakulása, 1945-1947." *Elméleti és Módszertani Közlemények*, 1970.

Index

and population exchange, 105, 106,
113
and preparation of peace conference,
44, 113, 114, 144, 145, 146
removed as foreign minister, 180
requests armistice, 24
signs armistice, 27
signs peace treaty, 174
telegram to peace conference, 164
Western trip, 133, 136, 137-138

Habsburgs, 28, 29, 113, 117, 152, 219,
241, 245, 278, 284
Hague Tribunal, 270, 275
Halifax, Lord, 76
Harriman, Averell, 24, 25, 53, 57, 62,
73, 96, 101
Harrison, Wallace, 256-257
Helm, Alexander Knox, 163-164, 266
Helsinki, 72
Herman, Alfred, 285
Hickerson, John, 136, 138, 146, 245-
246
historic kingdom, 30
Hitler, Adolf, 2, 22, 30, 31, 32, 50, 67,
102, 103, 113, 114, 142, 143, 155,
175, 217, 271
Hlinka Guard, 104
Holocaust, 67
Hopkins, Harry, 75
Horváth, Imre, 258, 288, 298
Horthy, Miklós, 35, 39, 43, 44, 46, 67,
80, 100, 113, 116, 123, 129, 138,
152, 171, 279, 295
Hungarian Communist Party (MKP),
(see also, Hungarian Workers'
Party [MDP]), 82, 175, 194, 197,
198, 213, 223, 255,279, 299, 306,
307
and approach to western democracies,
132
and conspiracy case, 202, 220
early policies of, 33, 107, 165, 166,
174, 195
and elections, 90- 91, 92-93, 190, 205
and fight over police control, 43, 135
foreign perceptions of, 124
formation of, 20
fuses with Social Democratic Party,
217

and Holocaust, 69
in Provisional National Assembly, 22
initial estimates of membership in, 22
peace plans of, 114, 144
and proclamation of republic, 116
and Red Army, 61
relations with Social Democratic
Party, 44, 92, 183, 210, 229
and revisionism, 128
and Yugoslav Communist Party, 222
Hungarian National Union, 116
Hungarian Unity Society, 167
Hungarian Workers' Party (MDP) (*see
also* Hungarian Communist Party
[MKP]), 217, 229, 230, 234, 253,
294
Husak, Gustav, 53, 232

Ignatiev, Yacov, 219
Imrédi, Béla, 214
Industry, 29, 232
International Monetary Fund, 96
Israel, 287
Istanbul, 8
Italy, 12fn, 28, 31, 76, 89, 113, 126,
156, 182, 187, 199, 200, 201, 213,
293

Jacobson, Israel, 266
Jebb, Gladwyn, 162
Jews, 22, 27, 38, 45, 66, 67, 68, 69, 104,
105, 137, 160, 175, 221, 232, 287,
307
JOINT Distribution Committee, 131,
266
Jones, Griffith, 271

Kádár, János, 254
Kállai, Gyula, 252, 256-257, 265-266,
268
Kállay, Miklós, 8, 9, 37
Kálló, Iván, 284-287
Kardelj, Edward, 200, 238
KGP (see Smallholders)
Kennan, George, 62
Kertész, István, 157
Key, William, 94, 105, 135
Kirk, Allen, 257
Koczak, Steven, 245
Korean War, 295-296, 298-299, 300, 302

143, 152, 173, 183, 212, 222, 267
anti-Semitism in, 68
borders with Hungary, 52, 89-90, 124, 126, 137, 139
charges of treaty violations by, 258, 270
and COMECON, 244
demands for elections in, 85, 95
Hungarian perceptions of, 282-285 passim, 287
in Little Entente, 3, 77
and Marshall Plan, 187
minorities after World War II, 102, 146
minorities between wars, 55
and Paris peace conference (1946), 154-155, 157-158, 160
and Paris peace settlements (1919-1920), 31
persecutions in, 56, 64
place in East Europe, 29, 128-129, 305
Soviets propose recognition of, 73, 75-76
surrender in World War II, 11, 13, 49
trade agreement with Hungary, 79
trade agreement with USSR, 81
and treaties with East European states, 175, 202, 204
and Vienna Award, 26
and Yugoslavia, 262
of republic, 116
Rusk, Dean, 246
Russia (see also USSR), 1, 17, 31, 40, 51, 68, 124, 127, 147, 149, 213, 226, 239, 308
Ruthenia, 22, 31, 50, 53, 80, 88, 103, 143

Sanders, Edgar, 265, 268-269, 271, 272, 274
Schoenfeld, Arthur, 40, 72, 73, 74, 90, 92, 94, 96, 99, 101, 111, 115, 123, 134, 151, 165, 166, 168, 176-177, 180, 278, 285
Schoenfeld, Rudolf, 285
Social Democratic Party (SDP), 14, 23, 37, 42-44, 83, 91, 93, 100, 111, 116, 132, 135, 144, 154, 172-174, 181, 203, 205, 210, 217, 220, 229, 230, 289

Sebestyén, Pál, 125-126
Second Hungarian Army, 32
Sik, Endre, 245-246, 256, 268, 288, 296, 298
Siroky, Jan, 231
Slansky, Rudolf, 199, 200, 283
Slavik, Juraj, 186
Slovakia, Slovaks (see also Czechoslovakia), 31, 50-51, 54, 68, 102-105, 112-113, 115, 126, 142-144, 152, 155, 159, 164, 200, 207, 230-232, 250
Slovak National Council, 53, 104, 112
Smallholders' Party (KGP), 83, 139, 188, 195, 222, 282
and commercial treaty with USSR, 101
communist view of, 132, 166
and conspiracy case, 167, 170-171, 177, 181
and elections, 93-94, 99-100, 165, 190
and land reform, 40
and peace terms, 113
and peace treaty, 174
and premiership, 180
in Provisional Government, 23
reactionism of, 116
and revisionism, 110
and Truman Doctrine, 172
and western powers, 42, 52
views on religion, 175-176
Smith, Walter Bedell, 143, 147, 157, 158, 211
Sofia, 72, 74, 213
South Africa, 156
South Slav Union, 222-223
Soviet Union (*see* Union of Soviet Socialist Republics)
Spain, 28
Squires, Leslie Albion, 72
Stalin, Josef V., 52fn, 153, 148, 155, 171, 198, 214, 218, 221, 243, 258, 296, 300, 301, 302
and Berlin Blockade, 218, 224, 226, 244, 294, 302
and China, 293
and Cominform, 199
his commanding position, 217, 220, 224
dealings with President Truman, 75
decline of Stalinism, 307-308
East European policies of, 2, 15, 16,